DARWIN *and Design*

MICHAEL RUSE

DOES EVOLUTION HAVE A PURPOSE?

HARVARD UNIVERSITY PRESS
CAMBRIDGE, MASSACHUSETTS
AND LONDON, ENGLAND 2003

Printed in the United States of America

Library of Congress Cataloging-in-Publication Data

Ruse, Michael.
Darwin and design : does evolution have a purpose? /
Michael Ruse.
p. cm.
Includes bibliographical references (p.).
ISBN 0-674-01023-X (alk. paper)
1. Evolution (Biology)—Philosophy. 2. Teleology.
I. Title.

QH360.5 R867 2003
576.8′01—dc21 2002038819

Designed by Gwen Nefsky Frankfeldt

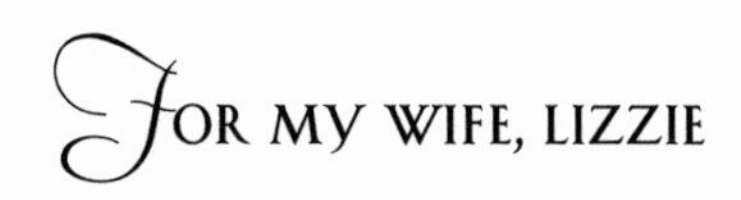
For my wife, Lizzie

Contents

Darwin and Design: Does Evolution Have a Purpose? is the third volume in what has now stretched to a trilogy, beginning with *Monad to Man: The Concept of Progress in Evolutionary Biology* and continuing with *Mystery of Mysteries: Is Evolution a Social Construction?* My aim through these three volumes has been to bring our understanding from several disciplines—philosophy, history, and religion—to answer questions about the nature of science, and conversely to use science to answer questions in the subjects I use as my probes. The underlying concern throughout is with the relationship between science and the culture that produces it, and if and how the two interact. I care about the question of values, of interests, and how and to what extent they appear in science and if and when they are reduced or eliminated through the course of time—or if, in some sense, science is always value-impregnated. Being a committed naturalist and believing that the way to solve problems is by looking at real-life issues, I have based my discussions on a specific case study: evolutionary thought, from its beginning in the eighteenth century down to the present day—a history where the key event was the publication of Charles Darwin's *On the Origin of Species* in 1859.

In *Monad to Man,* I followed the symbiotic relationship between evolutionary ideas and the cultural concept of progress. I wanted to see why so many have been attracted to evolutionary thought and why, even to this day, so many find it disturbing, if not outright upsetting, especially in its Darwinian version. In *Mystery of Mysteries,* reversing the approach, I used evolution's history to throw light on the much-discussed question about whether science is a disinterested reflection of objective reality or merely a subjective epiphenomenon of culture, a social construction. In *Darwin and Design* I go back to the beginnings of Western civilization and look at a powerful mode of thinking—teleological thinking—that has traditionally had important implications not just for our understanding of the world we inhabit but also for notions of a creator or deity. Darwinism is generally believed to have disturbed if not destroyed this traditional mode of thought, and my aim is to see if indeed this is true.

In recent years, a number of philosophically minded thinkers have raised the possibility that the nature of the physical world is such as to demand understanding in terms of teleology and deity. Enthusiasts for the so-called anthropic principle argue for a spectrum of claims, from the relatively innocuous observation that any world that supports life must be a world capable of supporting life to the much stronger conclusion that the world in which we live—a world that supports life—is so unlikely that its existence can be no mere coincidence and hence must be evidence of powers beyond nature. There are at least as many critics as there are supporters of these kinds of claims. In this book I have avoided entirely any discussion of these various topics and confined references in the physical sciences exclusively to those points at which they impinge upon the biological. Although I am not averse to controversy, my topic through these three books has been evolutionary theory, and now is not the time to broaden my scope. To do full justice to anthropic questions would take us into aspects of modern science that go quite beyond the physical sciences in their historical context as well as outside the issues of teleology as they are relevant to modern biology. I am confident that those whose main interest is in such questions will readily find that any conclusions I reach here can be reinterpreted in terms that they find pertinent.

In all of my three books, I hope that my findings are of interest in themselves and, at a more abstract level, shed light on the overall relationship between science and culture. While I would like to think that the volumes are sufficiently stimulating that readers will want to go from one to the next, I have written them so that they can be read independently. I have had a terrific amount of fun doing so, and I have learned a lot in the process. Evolution is a wonderful idea, and evolutionists (and their critics) are fascinating people. It has been a joy and a privilege to spend time with them.

INTRODUCTION

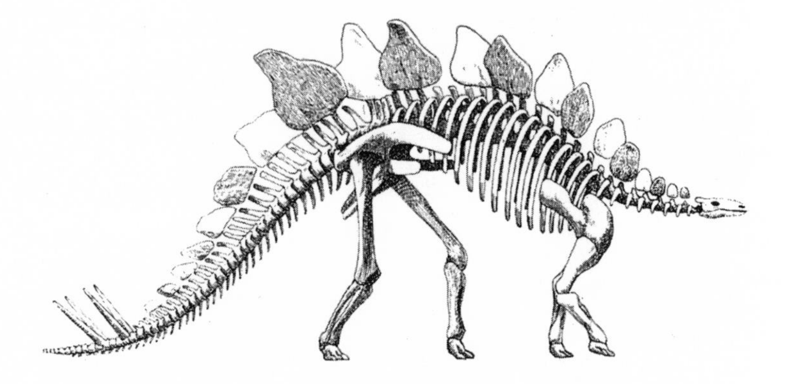

"KNOWLEDGE IS THE OBJECT OF OUR INQUIRY, AND MEN DO NOT THINK THEY KNOW A THING TILL THEY HAVE GRASPED THE 'WHY' OF IT (WHICH IS TO GRASP ITS PRIMARY CAUSE)."

THESE WORDS by the great Greek philosopher Aristotle hold as true today as they did in the fourth century BC. In trying to understand the world around and within us, we try to find the things or events that make other things or events happen: we try to find the *causes.* In trying to understand the rise to power of Adolf Hitler in 1933, for example, we look for the causes behind this event: the humiliating terms imposed by the victors on the defeated at the Versailles conference in 1919, the distaste felt by many powerful Germans for the democracy of the Weimar Republic, the crushing effects of the economic slump at the end of the 1920s, the personal fanaticism of Hitler himself, and more. All of these circumstances and others have been suggested as causes behind the triumph of the Nazis.

Now note something obvious but significant about these putative causes. They all come before the event being explained. The Versailles treaty was signed some fourteen years before Hitler became the German chancellor. The distaste for democracy comes at least as early, if not earlier. And so on for the other causes. Causes do not ever come after effects. What accounts for this asymmetry?

There is one simple but important explanation, often referred to as the problem of the "missing goal object." If causes come before effects, then you are never stuck with having to explain effects that have no causes. Versailles occurred and then Hitler occurred. If effects could come before causes, you would explain Versailles (the "effect" in this case) as being caused by Hitler's rise to the position of chancellor. But suppose that Hitler had been killed in the Munich putsch attempt of 1923. Then he would never have become chancellor, and Versailles would therefore have no cause. Or worse, you would have in theory the possibility of two identical Versailles treaties ("effects"), with identical events leading up to them, and yet with radically different causes. I am not sure that this is a logically impossible situation, but it is certainly very inconvenient from the viewpoint of understanding.

Causes come before effects. But do they always? Take the case of a graduate student trying to get a doctoral degree. She has to take a number of courses and examinations and then write and defend a thesis. Her aim in life is getting that degree. Here surely we could say that the future is reaching back and influencing the present. The degree that she will receive in the future is the cause of her taking courses today. Yet can this possibly be so? All too often, students drop out and give up. Or they fail their exams and get kicked out of the program. Our student comes to grief on the French qualifying test, and her plans are dashed to the ground. What can we say here about the future causing events in the past?

A solution is to regroup and argue that the cause of her actions today is not the actual fact of getting the degree in the future (which may or may not come to fruition) but rather her *intention* or hope or desire to get that degree. These psychological motivations may make reference to an imagined future, but they exist today. Hence, even the student who fails her French exam had the desire to pass and to go on to the thesis stage and ultimately to get her degree. The desires may well be the same for both the successful and for the unsuccessful student. Given the vagaries of life, no one ever said that identical causes had to have the same end.

In fact, this forward-looking kind of understanding, where future outcomes apparently influence events in the present, extends out from human desires to the artifacts that humans produce. My family and I enjoy fresh bread from the local bakery, but when we try to use a regular knife to cut a loaf that is still warm and steaming, the result is not so good. The bread is much too squishy. But we have a solution, namely a serrated knife that can cut fresh loaves with a sawing motion. Here too we can see a reference to the future: I purchase a knife now in order to cut bread in the future. The knife as it exists now is to be explained or understood by reference to bread-cutting still to come. Even if the knife snaps before use, there is no missing goal object. My past and present intention, transferred to the knife, caused it to be purchased in its serrated form, and this holds true even if the knife is never used as it was intended. The *purpose* or function or end of the knife was to cut soft, rubbery objects. It was made *in order to* slice fresh bread. That was the intention of its maker.

Here now comes the problem, or at least the background to our problem. Putting aside humans and their artifacts, the wonder is that forward-looking understanding—what philosophers call *teleological* understanding—seems to occur and to be appropriate in situations where there is no human intention whatsoever. Take the stegosaurus, a dinosaur remarkable chiefly because of the series of plates that alternate from side to side all the way down its spine. Why does the stegosaurus have these plates? What caused them? At one level, we might say that their cause was the embryological development of the animal itself. Various proteins were produced from the nucleic acids which carried the genetic information, and then from these the plates were constructed. But at another level, one's interest is forward-looking. What were the plates *for?* For what function did the plates come into being?

A number of answers have been proposed, and we will dig into them later. The popular answer today is that the plates aided heat regulation in this cold-blooded dinosaur. The stegosaurus had plates in order to control its temperature more efficiently. Here, where no human intention intervened, it still seems to make sense to us to invoke forward-looking language—language that explains the past by reference to a future outcome. The plates came into existence for the purpose of heat regulation.

The stegosaurus's plates are representative of a huge class of attributes in organisms that biologists attribute to adaptation. Adaptations serve the well-being of their possessors; they serve a purpose. We humans have many adaptations: eyes, noses, ears, teeth, penises, vaginas. And other organisms are likewise well-endowed. The little finches that Darwin studied on the Galápagos, for instance, have a wonderful variety of beak shapes and sizes, all of which are well-suited to the particular foodstuffs eaten by their owners. Finches with very thick strong beaks live on cactus and nuts and berries. Other finches with very fine, delicate beaks are exclusively insectivorous. Yet other groups with different-sized and shaped beaks eat a variety of foodstuffs. There are even species with beaks good for picking up and carrying sticks. They poke about in the bark of trees and shrubs for hidden insects and grubs.

Plants too have adaptations. Think of the leaves of the maple tree, beautiful to our eyes but essential to its owner for catching sunlight and using that energy to produce food. The leaves serve a vital purpose in the life cycle of the tree. Unfortunately, however, in all of these cases the missing-goal-object problem once again raises its ugly head. What if the dinosaur came to grief before it had time to make use of its plates? What if a man becomes a priest and never uses his genitalia for reproduction? What if there is a drought on the Galápagos and all the birds starve to death, regardless of what kind of beaks they have? What if acid rain kills all of the maples?

The traditional answer is that this all goes to show that we humans are not the only beings with intentions. Adaptations are God's handiwork, fashioned to help organisms (including humans) to function down here on earth. The plates that were never used, the genitalia that went unexercised, the beaks whose owners fell lifeless to the ground, the leaves that withered before their time were nevertheless fashioned by God with future purposes or circumstances in mind. We have here one of the standard arguments for God's existence, the so-called *argument from design,* or the teleological argument. The eye (to take the favorite example) is analogous to a telescope: telescopes have telescope-makers; hence, eyes must have eye-makers. We humans have our intentions and purposes, and so God likewise has his.

Or so it went until 1859, when the English biologist Charles Robert Darwin published *On the Origin of Species by Means of Natural Selection.* In this volume, he argued that all organisms, including humans, are the end-product of a long, slow, lawbound (natural) process of evolutionary change, up from "one or a few forms." And he proposed a mechanism which has brought that change about: natural selection, also known as the survival of the fittest. It works like this: Since more organisms are born than can survive and reproduce, a struggle for existence ensues, and success in that struggle is a function of the features that the successful organisms possess. This natural process is akin to artificial selection practiced by breeders, who pick from each litter the pups that possess the characteristics desired. Given enough time, natural selection, too, results in full-blooded change. The end result is a vast diversity of life, marked distinctively by the fact that each characteristic of species works, it functions, it is adaptive.

Darwin was most certainly right, and his theory seemed to sound the death knell for a forward-looking way of regarding the living world. Already, few people thought of the inorganic world—the stars, planets, mountains, oceans—as having a purpose that provides evidence of the Creator's design. Rarely does anyone ask what function the moon serves (other than wags who claim that its purpose is to light the way home for drunken philosophers). No one says that Niagara Falls exist in order to serve some particular end. It is true that the Falls (the Canadian Falls, that is) produces a rainbow-like spray, and a lot of honeymoon-hotel operators and others make money out of this. But the Falls does not exist in order to produce the rainbow, and certainly not to line the pockets of the local hotel franchise owner.

Analogously, after Darwin, one anticipates that biologists would consider it wrong to speak of the plates on the stegosaurus as existing in order to produce heat regulation. It may well be that the plates do regulate heat, just as Niagara Falls produces the rainbow. But after the *Origin*, most biologists no longer tried to explain or understand the regulatory function of the plates in terms of God's intentions, even if they believed that he exists. To them, the living world is just as much a natural, law-bound—that is, nonsupernatural—domain as is the nonliving world. Function-talk in biology became otiose at best, misleading at worst.

Yet, Darwin or no Darwin, we all—biologists included—still use function-talk. We speak of the purpose of the stegosaurus's plates and the function of the eye or the hand. We still say that the finch has a strong stubby beak in order to crack nuts. In other words, we still use the language that theists used in Darwin's day. That most Darwinian of biologists, Richard Dawkins—a brilliant science writer and fervent atheist—positively boasts of his commitment to the teleological way of thought. At the beginning of *The Blind Watchmaker* (1986), one of his most popular books, Dawkins speaks of his conviction that "our own existence once presented the greatest of all mysteries," one made the more surprising because so many people are "in many places actually unaware that there was a problem in the first place!" And what is this mystery? "The problem is that of complex design" (p. ix). Nor is Dawkins alone in this belief, or in using such language. In one of the most highly regarded books by a still-living evolutionist, the American ichthyologist George Williams (a man who jokes that the only good advice he got from his parish priest was the direction to the front door of the church) is explicit beyond doubt: "Whenever I believe that an effect is produced as the function of an adaptation perfected by natural selection to serve that function, I will use terms appropriate to human artifice and conscious design. The designation of something as the means or mechanism for a certain goal or function or purpose will imply that the machinery involved was fashioned by selection for the goal attributed to it" (Williams 1966, 9).

This then is the paradox to which *Darwin and Design* is directed. Darwin seems to have expelled design from biology, and yet we still go on using and seemingly needing this way of thinking. We still talk in terms appropriate to conscious intention, whether or not we believe in God. In biology, we still use forward-looking language of a kind that would not be deemed appropriate in physics or chemistry. Why is this? And what does it say about the way we humans think? As an evolutionist, I believe the answers to questions about the present are to be found in the past, and this principle will be my guide here. I will look first at the history of forward-looking or teleological language in the pre-Darwinian era and then go on to examine Darwin's own thinking, both the general position he promoted and the specific implications for our study. I will also see how things fared in the post-Darwinian era, at the hands of both supporters and critics of evolution through natural selection.

From this base, we will be in a position to ask where design-like thinking stands today, going from the ideas of active, working evolutionists, through the thinking of philosophers of science, and on to the beliefs, hopes, and arguments of present-day theologians, who may or may not be fighting a rearguard action in the face of the *Origin*. I think our paradox has a solution, which I shall offer. But more than this, I believe that the question of design opens an important and fascinating window onto the natural world and our thoughts about it, and onto the relationship between the two.

CHAPTER 1

TWO THOUSAND YEARS OF DESIGN

ATHENS AND JERUSALEM. The Greeks and the Jews. So much of Western culture can be traced to these two ancient civilizations, and this is true of our tale about evolution and design. We look back to the Greeks for the flowering of reason—of rational thought, philosophy, mathematics, science—and we look back to the Jews for the beginnings of faith—of revelation, monotheism, devotion, and feeling. Consider the great Greek moralist Socrates (fifth century BC) and the Jew who founded Christianity, Jesus, both of whom fell afoul of the authorities and were executed; both of whom went to their deaths refusing to recant or to apologize or to resist or flee. Yet the differences were stark. Socrates was arguing about God and immortality right up to the hour of his execution. Jesus, by contrast, was never much given to open discussion. During his life, he spoke in parables and issued commands for proper belief and behavior; and at the end, the resignation he exhibited on the Cross represented the ultimate act of faith and acceptance. Of course, one should be wary of generalizations. The ancient Greeks were hardly indifferent to the power of emotions (one thinks of Euripides), and the Jews were hardly beyond the reach of reason (one thinks of Saint Paul). Still, different attitudes, different methods, different goals are apparent in these two great systems of thought.

Our story begins on the Greek side, in the fifth century BC, with the school of philosophers known as the atomists. Led by Leucippus early in the century and then somewhat later by Democritus, these thinkers argued that the whole of reality consists of hard little particles (atoms) forever rushing around in infinitely empty space (the void). Nothing exists, nothing is real, save these particles, which have only size and shape; colors and tastes and all else are secondary qualities invented by the perceiver. The particles move randomly, and given enough time and space—unlimited amounts of both—sometimes they collide and cohere. Whole universes form quite by chance in this way, and within these universes smaller bodies—living bodies—eventually come into being. The atomists thought that bits and pieces of organisms would be formed first, and these in turn would collide and cohere. Some formations, those that made functioning organisms, would survive and persist. Those with incomplete or mismatched or misplaced pieces—an eye here, a nose there, three legs, no anus—would not. This all accorded with natural law as the atomists understood it—blind law, that is, pure chance without plan.

Of course, this was all crazy talk, at least in the eyes of the greatest of Greek philosophers, Socrates' student Plato (427–347 BC), and also in the eyes of Plato's own student Aristotle (384–322 BC). For these two men, even with infinite time and infinite space, inert matter moving randomly could not make a functioning universe. But what were the alternatives? Plato and Aristotle provided their own answers—answers with different emphases—that were to persist through the ages until Darwin, and perhaps even down to our own time.

Plato's Argument from Design

Plato expressed his ideas in the form of dialogues, many purporting to be conversations between Socrates and his young male followers, aristocratic Athenians who would gather at Socrates' feet to listen, debate, and learn. The early dialogues are probably authentically Socratic—fairly accurate reports of actual discussions between Socrates and his followers and opponents. But at some point, although Plato still used the dialogue form, he started to insert his own ideas. Socrates himself probably formulated a version of the argument, but it was Plato's distinctive thinking on the problem of design in these later dialogues that was to influence posterity.

First some crucial background on Plato's beliefs about the nature of existence or being. Plato's ontological theory centers on the eternal Forms: patterns or templates representing universal paradigms, of which the things of this world are mere temporal copies or reflections. Belonging to some metaworld of ultimate rationality, shared only with the laws of mathematics, the Forms reach the peak of reality in the Form of the Good, which gives life and illumination. In our world, this Form is represented by the Sun. Since our domain of experience is only partly real—real only inasmuch as it "participates" in the world of the Forms—ours is the domain of becoming and decay, of change and time, of wrong as well as right. Yet despite these flaws, this domain of ours is not one of mere chance and chaos—we have order, as seen in the motions of the heavenly bodies. In the Platonic heavens (in some sense the true abode of human souls, which are themselves eternal), the stars forever trace their paths and patterns as they move in perfect, endless circles.

In his great dialogue the *Phaedo,* which tells of Socrates' death, Plato drew a sharp distinction between causes that simply act without intention (what we would call efficient causes) and those that act purposefully, according to plan, with some end result in mind. Consider the question of why a man grows. "I had formerly thought that it was clear to everyone that he grew through eating and drinking; that when, through food, new flesh and bones came into being to supplement the old, and thus in the same way each kind of thing was supplemented by new substances proper to it, only then did the mass which was small become large, and in the same way the small man big" (*Phaedo* 96d). But then, Plato (using Socrates as his mouthpiece) continued to say that this kind of physiological explanation of growth, unaided, will not do. It is not wrong, but it is incomplete.

As Plato saw clearly, the real key to understanding this kind of causation is value and purpose: things that someone or some people want and take steps to attain. "The ordering Mind ordered everything and placed each thing severally as it was best that it should be; so that if anyone wanted to discover the cause of anything, how it came into being or perished or existed, he simply needed to discover what kind of existence was *best* for it, or what it was best that it should do or have done to it" (97b–c). Socrates sitting in a jail cell cannot be explained as a matter of blind chance, as simply the workings of the laws of physiology on his muscles and sinews and so forth. We must get into questions about intention, about why he did not escape when he might have, for example. "If someone had said that without bones and sinews and all such things, I should not have been able to do what I decided, he would be right, but surely to say that they are the cause of what I do, and not that I have chosen the best course, even though I act with my mind, is to speak very lazily and carelessly" (*Phaedo* 99a–b). We understand some things in terms of their inherent value to us, and we see other things as making possible (causing) the successful realization of our values in the future. I want to get a PhD at some later date (I value it), and so I strive to pass the language examination (which I now also value). I want a piece of fresh bread properly sliced, and so I buy a serrated knife.

In Plato's example of growth, eating and drinking cause future growth and development, but because these are things we prize, it is proper to go further and say that eating exists *for the purpose of* bringing about growth and development. If we did not want to grow and develop and were content to stay six forever and ever, then we would not speak and think in terms of the *purpose* of eating. It would be enough to give the first kind of explanation about how eating and drinking cause growth. There would be no more call for further explanation than there is when we say that stubbing your toe causes the nail to turn black and fall off. You did not stub your toe *so that* the nail would fall off. You stubbed. It fell. That is all.

According to Plato, this is the problem we face when we try to explain causes. It may seem that we are facing a funny kind of causation, where cause and effect are reversed in time. The cause is the thing in the future, bringing about the effect in the present or past. The growth brings on the eating and drinking, and so forth. But this is not the essence of the kind of causation that we are considering now. Values are what really count, and where values come into play—things that people desire—we need a different kind of explanation, an explanation that refers to the ends and purposes of things. Wants and desires imply consciousness, intention; values imply a mind. The stone in the brook does not value the water flowing over it, but I do value growing big and strong.

Here we reach the controversial part: Whose mind is it that puts everything in motion and orders things for their own future good? It is hardly our own minds—at least, it is hardly our minds once we look beyond our intentions and desires. *We* did not decree that eating would be of importance in achieving growth and maturity. It is rather, according to

Plato, the Mind of God. Not necessarily the Jewish or Christian God, in the sense of one who created a world from nothing. But certainly a designer God, whose nature Plato nailed down in his later dialogue about the workings of the universe, the *Timaeus.* The detailed speculations of this work may or may not have been endorsed by Plato himself, but its general philosophy was surely congenial.

The *Timaeus* draws a distinction between causes that just make things happen and causes that give real understanding. "Both kinds of causes should be acknowledged by us, but a distinction should be made between those which are endowed with mind and are the workers of things fair and good, and those which are deprived of intelligence and always produce chance effects without order or design" (*Timaeus,* 46d–e). Plato identifies this mind with a creator, what he called the Demiurge. It is not necessarily one who makes something out of nothing (a god who creates *ex nihilo*) but an orderer of this world belonging to and guided by the world of rationality, the world of the Forms and of the Good.

What sorts of things demand explanation in terms of ends, values, or purpose? Plato gives many examples drawn from life, but explanations that invoke purpose are in order for the inanimate world as well, for it too is an object of design. In a way, Plato saw the whole universe as a kind of superorganism, with its own soul. Be this as it may, whether he is talking about the animate or inanimate world, Plato's argument for an intelligent mind in creation has two major moves. The first move is to prove that there is something about the world which needs explaining, which simply cannot be attributed to blind chance. Human growth and the universe as a whole are not directionless and disordered. Their structure and operation show a high level of order. This first Platonic move is often called the *argument to design,* because it gets us from our starting point to a recognition that the world exhibits design. In this leg of the argument Plato is trying to establish that design exists to be explained.

I confess that I am uncomfortable with this terminology in this context, because it seems to prejudge a basic question being addressed in this book, which is: Does the world actually exhibit design? Perhaps it does, or perhaps it doesn't. Or perhaps some of it does and some of it doesn't; one does not want to be forced into judging everything in black and white—either all is design or nothing is. One might conclude that organic adaptation calls for special attention but that the general laws of mechanics and their effects do not. One might substitute for the term *argument to design* the alternative term *argument to order,* but this is not quite strong enough. The periodic table is ordered, but whether it has the feature we are trying to pick out is a debatable point. Presumably Plato, believing that the inorganic shows design—remember that for him, ultimately, the organic and the inorganic collapse into one—might have argued that the periodic table, had he known of it, shows the desired quality. But even he seems to agree that not all natural phenomena have quite what we are trying to identify here.

Another term, the *argument to complexity,* founders, I think, on the various meanings of complexity. A thunderstorm might be thought complex, and yet it calls for no special understanding. Richard Dawkins (1983) speaks in terms of *organized complexity* or *adaptive complexity,* yet even here we are hardly getting away from anthropomorphic terms. Who did the organizing of organized complexity? Perhaps we would do better to say *apparent organized complexity,* and probably this is about as good as we are going to do. Apparent organized complexity or *adaptive complexity* can be abbreviated simply to *order* or *complexity,* so long as the context is understood. The important point here is that the first move in Plato's argument is to recognize the distinctive nature—I would call it the complexity—of certain things whose existence requires some kind of special explanation.

After the argument to complexity (the argument that moves us to a recognition that complexity exists), Plato's second move is from the complex, distinctive nature of things to an explanation of that distinctive nature. This is sometimes referred to as the *argument from design*—the move from acknowledgment of design toward acknowledgment of a designer of some sort. I am even more uncomfortable with this second term, because the move itself strikes me as almost trivial. If we all agree that literal design has occurred, there must, by definition, be a designer (although the nature of that designer is, of course, another matter). But has design actually occurred, just because complexity exists? That is the question that concerns us in this book—whether, having made the move to complexity, one is committed to making the subsequent move to design.

I prefer to retrieve the term *argument to design* for this second move, meaning an argument that gets us from recognizing the distinctive complexity of the world to acknowledging that this complexity entails design (and, by definition, a designer of some sort). Can one block this second move, either by blocking the first or by allowing the first but cutting off the second? That is, can we argue convincingly that there is no complexity? If not, can we argue convincingly that complexity exists, but that it does not imply design?

The second part of the argument, the move from the complexity of the world to the mind behind it—the designing mind—was Plato's true obsession. He spent little time on the first part of the argument, the part that looks at the distinctive nature of the world and what makes it distinctive. Purpose in Plato's teleology was an *external* purpose—imposed upon things by outside factors, in this case the Designer God.

Aristotle's Final Causes

Aristotle took so much from Plato, and yet he differed from him on many points—nowhere more so than over ideas about design and purpose in the universe. Aristotle's views came as part of his overall analysis of causation. He claimed that there are four different senses in which phenomena can be said to bring about, or to cause, or to be the causes of, other phenomena. The one of interest to us is the fourth, *final causes*, where things occur for the sake of desired goals. In Aristotle's *Physics*, just as in Plato, we find human intentionality: we ourselves do something, or we make instruments to do something, with our own ends in view. But Aristotle was a practicing biologist for part of his life, and, although he used a human model to explain what he was about, in his biological discussions he introduced final causes without direct reference to intentionality at all.

Criticizing the atomists' belief that everything just happens by chance and that there is no need for end-related thinking, Aristotle drew an analogy between a piece of furniture and the features, or characteristics, of an organism. Just as we think of furniture as made by certain actions of a craftsman for some end, so also we should think of the organic part as made by certain actions for some end. Distinguishing a model hand from a real hand, and criticizing physiologists who think that all they need to do is to refer to the immediate causes of features, Aristotle chided: "What are the forces by which the hand or the body was fashioned into its shape?" A woodcarver (speaking of a model) might say that it was made as it is by an axe or an auger. But simply referring to the tools and their effects is not enough. One must bring in ends. The woodcarver "must state the reasons why he struck his blow in such a way as to effect this, and for the sake of what he did so; namely, that the piece of wood should develop eventually into this or that shape." Likewise against the physiologists, "The true method is to state what the characters are that distinguish the animal—to explain what it is and what are its qualities—and to deal after the same fashion with its several parts; in fact, to proceed in exactly the same way as we should do, were we dealing with the form of a couch" (*Parts of Animals,* 641a7–17).

We see that Aristotle had in mind the model of a craftsman, as did Plato, but his was not an argument intended to prove a designing mind. For him, the end direction, the purpose, was more naturalistic—it is more part of the way that nature works. His emphasis was on the argument to complexity rather than the argument to design. He stressed the end-directed nature of the things under discussion. But what about the Designer, the obsession of Plato? What about the argument to design? Does it make sense to talk of interests or wants without a consciousness behind them? Can we have values without a valuer? Clearly, in a sense, Aristotle thought we can. The good or the final cause is the well-being of the individual organism. An organic feature exists for its possessor's well-being. It has its value in this sense, rather than in any value desired and imposed by an external being. An organic feature is part of the nature of things. It is constitutive of objects. It belongs to their ontology. In this sense, Aristotle's explanation of final causes was *internal* as opposed to the external teleology of Plato.

It is for this reason also that, despite his general discussion in the *Physics,* Aristotle's philosophy of purpose and design focused much more directly on the world of organisms. Indeed, some have argued that his teleology was focused exclusively on organisms. Whereas Plato saw a purpose in the whole universe, Aristotle worked at the individual, physical level. Inanimate objects seem not to have a purpose or an end—in the *Meteorology,* for instance, there is no mention whatsoever of final causes. Organisms, however, do have ends or final causes in his system.

Not that Aristotle was entirely naturalistic in the sense of thinking that final causes have no metaphysical underpinnings. You cannot simply drop consciousness out of the picture and then believe that value-based thinking remains untouched. Ultimately, Aristotle thought that the overall picture is one of parts functioning for the sake of wholes (organisms), which in turn are able to flourish and reproduce and thus participate in the eternal, which for Aristotle is identified with the deity or the divine. None of this is terribly helpful or consoling, however, because Aristotle's ultimate deity—the Unmoved Mover—has no knowledge of us and spends its time contemplating its own perfection. What *is* helpful and consoling is that Aristotle has put in place another part of the puzzle about purpose and design. Plato had explicated the value component and had seen that this ties in somehow with mind and forethought. Aristotle—a hands-on biologist as well as a philosopher—saw that a designing mind as such can have no place in science. Yet, we cannot do without purpose-like understanding, at least in the realm of organisms. In some way, this purpose or design in the living world must be understood right down at the level of the individual, with all its adaptations.

Aristotle left much labor for others to do in reconciling the external and the internal, but the groundwork was laid. More provocatively, one might say that the paradox was proposed. Thinking in terms of ends means thinking in terms of values, and values imply a consciousness. Yet science has no place for such a consciousness. What is one to do?

The Christians

As we leave Athens for Jerusalem, we must move the clock forward, past the birth of Jesus and into the Christian era. But as we do, we should note that Greek thought did not die or go absent. Indeed, as the power center of the ancient world moved west toward Rome, we find that final-cause thinking—both the argument to complexity and the argument to design, which taken together I prefer to call the *argument from design*—was picked up, cherished, and elaborated. The great Latin orator Cicero (106–43 BC) argued in his *De Natura Deorum* for the intricate nature of the living world, for the artifact or craftsman-produced nature of things like the eye, and concluded that such end- or purpose-like complexity could not have come about blindly or by chance. Likewise, two centuries later, the highly influential anatomist and physiologist Galen (129–c. 200 AD) took an explicitly Aristotelian attitude toward the living world. "Aristotle is right when he maintains that all animals have been fitly equipped with the best possible bodies, and he attempts to point out the skill employed in the construction of each one" (Galen 1968, 1.108). Following his own prescription, Galen looked at all anatomical parts from a final-cause perspective. The hand, for instance, has fingers because "if the hand remained undivided, it would lay hold only on the things in contact with it that were of the same size that it happened to be itself, whereas, being subdivided into many members, it could easily grasp masses much larger than itself, and fasten accurately upon the smallest objects" (Galen 1968, 1.72).

What about the Christian thinkers? Conveniently for our purposes, we can focus on their contributions for the first millennium and a half after Christ, breaking with them only when we reach the sixteenth century, with its Renaissance flowering of art and literature (much of it reaching back to the Greeks), the Reformation of the Catholic Church and the rise of Protestantism, and the Scientific Revolution, when Copernicus put the sun at the center of the universe and a host of other scientists refined and extended his discoveries.

The greatest theologians of the earlier era—the first fifteen hundred years of Christianity—were practitioners of a religion that grew out of Judaism. Yet, they were much influenced by the Greeks and struggled to synthesize the reasoned conclusions of the philosophers with the revealed truths of their Judeo-Christian faith. First there was Saint Augustine (354–430), much taken with the philosophy of Plato, and then eight centuries later there was Saint Thomas Aquinas (1225–1274), who devoted his life to reconciling Christianity with the newly discovered works of Aristotle.

Both Augustine and Aquinas saw in the insights of the ancients a way toward an appreciation of the existence and nature of the Christian God. For both of these men the argument to design was the really important idea. The argument to complexity left them relatively unmoved; they

were not practicing scientists, as Aristotle had been, or close to science and mathematics, as was Plato. They were theologians first and philosophers second. Thus, teasing apart the intricacies of the empirical world held little fascination. Whatever high rank these men gave "natural" theology (that is, reason-based argumentation for a Divine Designer), it necessarily played a secondary role to "revealed" understanding or belief. For Plato, the argument to design stood on its own; it led one to a belief in a god or gods. For Augustine and Aquinas, Christian revelation, particularly as presented in the Holy Bible, was the essential starting point for the believer.

Augustine was unambiguous on this point: "Of all visible things, the world is the greatest; of all invisible things, the greatest is God." But how do we know this, since we see the world but can only believe in God? The answer is simple: "That God made the world, we can believe from no one more securely than from God Himself. Where have we heard Him? Nowhere more clearly than in the Holy Scriptures, where his prophet said: 'In the beginning God created the heavens and the earth'" (1998, 452). It is against this background—Augustine and Aquinas knowing already that God was Creator of heaven and earth and that his mark would be on his creation—that the design argument makes its appearance.

First Augustine, in his greatest work, *The City of God*. "The world itself, by the perfect order of its changes and motions, and by the great beauty of all things visible, proclaims by a kind of silent testimony of its own both that it has been created, and also that it could not have been made other than by a God ineffable and invisible in greatness, and ineffable and invisible in beauty" (pp. 452–453). This is it. A part of a paragraph tucked in among hundreds of pages on other topics. Augustine was brief on this subject not because he thought the argument a bad or weak one. After all, it shows the existence of God and demonstrates some of his major attributes—he is powerful and has a sense of beauty and so forth. Rather, Augustine raced past this topic because, for him, the real theological action lies elsewhere: in faith and revelation.

The same is true of Aquinas, although he was more concerned than Augustine with collating revelation and reason. For him, the real force of the argument from design—one of five famous proofs that he gave for the existence of God—was not really that it proved God's existence in the face of unbelief. Aquinas did not live in a society of skeptics and agnostics and atheists. Rather, as one who had argued strenuously against the argument that the very idea of God proves necessarily that God exists (the so-called ontological argument), Aquinas now needed the design argument to show that one could nevertheless reach a knowledge of God's existence through reason.

But not reason unaided. Reason that starts with the world of experience—experience of the entire world, not just the world of organisms. "The fifth way is taken from the governance of the world. We see that things that lack intelligence, such as natural bodies, act for an end, and this is evident from their acting always, or nearly always, in the same way, so as to obtain the best result. Hence it is plain that not fortuitously, but designedly do they achieve their end." Then from this premise (the argument to order), more claimed than defended, he moves to the Creator behind things (the argument to design). "Now whatever lacks knowledge cannot move towards an end, unless it be directed by some being endowed with knowledge and intelligence; as the arrow is shot to its mark by the archer. Therefore some intelligent being exists by which all natural things are directed to their end; and this being we call God" (Aquinas 1952, 26–27).

As a good Christian, Aquinas, like Augustine, would never have claimed that we humans can have full and complete knowledge of God. He is infinite and perfect; we are finite and tainted with sin. God is cloaked in mystery; "we see through a glass, darkly." But we must have some understanding of God, more than just the promise that in the future we shall see "face to face." Otherwise, the very act of worship becomes unfocused and worthless. Answering this need, Aquinas worked out the theory that our understanding of the Deity and his attributes is analogical rather than direct. We can properly speak of God as a "father," not because he was the biological being who produced the sperm that fertilized our mothers but because he has father-like attributes—he cares for us and worries about us when we are troubled, as well as having created us all, of course. The argument to design was an important piece in this picture, not because it proved the existence of God as such—that was not needed—but because it helped to flesh out our analogical understanding of God's attributes and powers. A world—a world such as we live in—gives clues to its creator and designer. We live in a world of order, of things that work according to their purpose. Hence, we can properly conclude that God is an intelligent being, one who ordered the universe with our welfare at heart.

A theology like this starts to introduce a kind of teleology that is absent from Greek thought. Aristotle's work focused on immediate ends: the needs of organisms and how nature provides for them. Aquinas introduces a historical dimension, stretching from a creation in which humans—made in the image of God but born in sin—are absolutely central, through God's own Incarnation in the person of Jesus Christ, who saves us from our sins, and on to the ultimate hope of eternal life. We have an eschatology, as we move from the daily needs and ends of organisms to the long-term plans for and ends of humans. In Aquinas' system, history itself has acquired a purpose. Its successful outcome has value—to God and to us humans. For now, we can ignore this historical dimension to purpose, but later it will come into our discussion.

The Scientific Revolution

The great Christian thinkers of late antiquity and the medieval period absorbed the Greek concerns with ends—the argument to complexity and especially the argument to design. During the intellectual and social upheavals that occurred in the middle of the last millennium, these issues resurfaced with explosive force. The Protestant Reformation emphasized direct access to God by ordinary people—and not just learned individuals. One finds God by faith alone *(sola fide),* and one is guided to him by scripture alone *(sola scriptura).* What need have we of the argument to design when it is enough to read the word of God and to open our hearts to his message? In a sense, reason can only get in the way of this direct channel.

Yet, although Reformation theologians were not warm and welcoming toward natural theology, they were not entirely unsympathetic to the exercise of reason, and its effect on natural theology was complex. As John Calvin in particular stressed, not even original sin can eradicate reason entirely or render it worthless, because reason is a gift of God (*Institutes* II.ii.12, in Calvin 1962). Grudging acceptance of natural theology was clearly reflected in the Protestant statements of faith ("Confessions"), because Saint Paul himself, no less, had endorsed some form of natural theological approach in his Epistle to the Romans—"Ever since the creation of the world his invisible nature, namely, his eternal power and deity, has been clearly perceived in the things that have been made" (Romans 1.20). Still, the Protestant reformers were never entirely comfortable with appeals to reason and wished to downgrade its significance as compared with the Bible itself. The "Belgic Confession," an early Calvinist statement of 1561, makes this clear: "He makes himself known to us more openly by his holy and divine Word, as much as we need in this life, for his glory and for the salvation of his own" (Müller 1903, 223).

Not just in the Protestant churches but in science also, enthusiasm for final causes started to decline. The world of Copernicus (who published his major work, *De Revolutionibus Orbium Caelestium,* in the year of his death, 1543) is a world of efficient cause, not final cause. We want to know what makes the planets go around the sun, not what purpose the endless cycling serves. The metaphor of the world as an organism went into steep decline, and a new metaphor—the world as a machine, unthinking, uneventful, simply going blindly through the motions—started to take over. In the mid-seventeenth century the British chemist and general man of science Robert Boyle was a major spokesman for this vision. But he was not alone, for his attitude fit nicely with the views of the methodologists describing and prescribing the new ways of doing science.

Francis Bacon (1561–1626) had led the attack on Greek teleology, wittily likening final causes to vestal virgins: dedicated to God but barren. He did not want to deny that God stands behind his design, but he did want to keep this kind of thinking out of science. The argument to complexity is not very useful in science, especially in the nonliving context, Bacon asserted. And whatever one might want to say about the argument to complexity for the living world, inferences from this to a mindful Designer have no place in science. "For the handling of final causes mixed with the rest in physical inquiries, hath intercepted the severe and diligent inquiry of all real and physical causes, and given men the occasion to stay upon these satisfactory and specious causes, to the great arrest and prejudice of further discovery" (Bacon 1605, 119). Mixing the physical with the biological, and clearly assuming that once you have established adaptive complexity then design follows as part of the package deal, Bacon continued that it is one thing *in philosophy* to speculate that the lashes on eyelids are to keep the eyes clean, or that skin is to protect animals from hot and cold. Or that bones are to keep animals upright without collapse, or that leaves protect the fruit on the tree, or even that the earth is solid so that animals and plants have somewhere to live. But the introduction of this kind of argument *into science,* he wrote, is "impertinent" (p. 119).

The French physicist, mathematician, and philosopher René Descartes (1596–1650) felt much the same way. Introducing an argument that was as much theological as philosophical or scientific, he warned that "when dealing with natural things, we will, then, never derive any explanations from the purposes which God or nature may have had in view when creating them and we shall entirely banish from our philosophy the search for final causes. For we should not be so arrogant as to suppose that we can share in God's plans." Stick rather to efficient causes, "starting from the divine attributes which by God's will we have some knowledge of, we shall see, with the aid of our God-given natural light, what conclusions should be drawn concerning those effects which are apparent to our senses" (Principle 28, *Principles of Philosophy;* in Descartes 1985, 202). Wary of the power of the Church as always (he was writing at the time of Galileo), Descartes warned that we should never presume to go against the truths of revelation. But ultimately, like Bacon, Descartes could see no place for the design argument in our scientific understanding of the world of experience. Leave it to the theologians and the philosophers to make the overall argument for design.

This attitude fit very well with a rekindled enthusiasm for atomism (or the corpuscular theory, as it came to be known). Even in Roman times, atomism had enjoyed a major revival, with a consequent downplaying of intention and design. The late Athenian philosopher Epicurus (c. 341–270 BC) had not only adopted the Democritan ontology—small particles in infinite space or void, churning about endlessly without purpose or intention—but had tied this to a general philosophy of life that stressed the distance and indifference of the gods and urged a life of contentment and moderation and (since there is nothing beyond) consequent lack of fear of death. In *De Rerum Natura* (*On the Nature of Things*), the Roman poet Lucretius (c. 95–52 BC) had penned 7,500 hexameter lines in praise of this vision (Lucretius 1969, 32–33):

> The nature of everything is dual—matter
> And void; or particles and space, wherein
> The former rest or move. We have our senses
> To tell us matter exists. Denying this,
> We cannot, searching after hidden things,
> Find any base of reason whatsoever.

So much for organized complexity. So much for minds or Mind, outside the material. So much for trying to make any ultimate sense of anything beyond the immediate.

The scientists of the seventeenth century were more sophisticated than the Roman poet, and people like Boyle truly tried to tie in their thinking about atoms with the physical and chemical experiments they were performing. But there was undoubtedly a tug away from meaning (of some sort) and toward blind, purposeless law and all that it entails.

David Hume

What then were theologians and philosophers to make of final causes? By the end of the eighteenth century the answer would seem to be "not very much." The Scotsman David Hume (1711–1776), once wittily described by a fellow countryman (David Brewster) as "God's greatest gift to the infidel," was an empiricist and a skeptic. He reduced all knowledge to sensation and then doubted that certain knowledge was ever attainable. The best consolation we have is that our psychology will not let us believe our philosophy, and so we are able to go about the business of our daily life. Sentiment and emotion drive us, and reason can never be anything other than the slave of the passions. Hume saw religion as little more than something to stave off the trials of life and fears of death and the unknown. He was himself no out-and-out atheist—absolute nonbelief, he thought, requires a commitment no less justified than Christianity—but he was deeply skeptical of any claims to belief.

What then of the arguments for the existence of God: what in particular of the design argument? Hume's *Dialogues Concerning Natural Religion* is the most sustained attack ever penned against theology and religious belief of any kind. First, through one of the participants in the encounter, Cleanthes, Hume sets up the argument in its classic form. He starts by arguing for the special nature of the world (the argument to complexity), and from this goes on to infer that something creative stands behind it (the argument to design).

> Look round the world: contemplate the whole and every part of it: You will find it to be nothing but one great machine, subdivided into an infinite number of lesser machines, which again admit of subdivisions, to a degree beyond what human senses and faculties can trace and explain. All these various machines, and even their most minute parts, are adjusted to each other with an accuracy, which ravishes into admiration all men, who have ever contemplated them. The curious adapting of means to ends, throughout all nature, resembles exactly, though it much exceeds, the productions of human contrivance; of human designs, thought, wisdom, and intelligence. Since therefore the effects resemble each other, we are led to infer, by all the rules of analogy, that the causes also resemble; and that the Author of Nature is somewhat similar to the mind of man; though possessed of much larger faculties, proportioned to the grandeur of the work, which he has executed. By this argument *a posteriori*, and by this argument alone, do we prove at once the existence of a Deity, and his similarity to human mind and intelligence. (Hume 1779, 115–116)

Then, through another of the participants, Philo, Hume knocks it all down like a house of cards. If we liken the world to a machine, he asks, then are we opening the way for multiple planners of machines and many prior less adequate worlds? "If we survey a ship, what an exalted idea must we form of the ingenuity of the carpenter, who framed so complicated, useful, and beautiful a machine? And what surprise must we feel, when we find him a stupid mechanic, who imitated others, and copied an art, which, through a long succession of ages, after multiplied trials, mistakes, corrections, deliberations, and controversies, had been gradually improving?" More generally: "Many worlds might have been botched and bungled, throughout an eternity, ere this system was struck out: much labour lost: many fruitless trials made: and a slow, but continued improvement carried on during infinite ages in the art of world-making. In such subjects, who can determine, where the truth; nay, who can conjecture where the probability, lies, amidst a great number of hypotheses which may be proposed, and a still greater number which may be imagined" (p. 140).

This is a counter to the second phase of the design argument, from complexity to a Creator, that we might want to take seriously. Hume also went after the argument to complexity itself, which sets out to prove that something special exists that is in need of explanation. In Hume's opinion, we should be careful about making any such inference. We might question whether the world really does have marks of organized, adaptive complexity. For instance, is it like a machine or is it more like an animal or a vegetable, in which case the whole argument collapses into some kind of circularity or regression? It is certainly true that we seem to have a balance of nature, with one part's change affecting and being compensated by a change in another part, just as we have in organisms. But this seems to imply a kind of non-Christian pantheism. "The world, therefore, I infer, is an animal, and the Deity is the SOUL of the world, actuating it, and actuated by it" (pp. 143–144).

And if this is not enough—going back again to the argument for a designer—there is the problem of evil. It belies the optimistic conclusions about the Deity behind the design drawn by design-argument enthusiasts from Socrates on down. If God designed and created the world, Hume asked, how do you account for all that is wrong within it? If God is all powerful, he could have prevented evil. If God is all loving, he would have prevented evil. Why then does evil exist? Speaking with some feeling for life in the eighteenth century, Hume asked meaningfully, "What racking pains, on the other hand, arise from gouts, gravels, megrims, tooth-aches, rheumatisms; where the injury to the animal-machinery is either small or incurable?" (p. 172). Not much "divine benevolence" displaying itself here, he implies.

Unneeded in science, riddled with paradox in philosophy, and obstructive to genuine belief in religion—proving, if anything, the existence of just the kind of god one does not want to have around—teleological thinking seemed destined for the slag heap of discredited ideas, along with phlogiston, which was also making a deserved exit at the end of the eighteenth century. No good case had been made for something special about the world, no need to infer a creative intelligence behind everything was generally felt, and no way was seen to infer it, even if a need existed. After a run of more than two thousand years, understanding in terms of final causes, purpose, or design seemed to have come to a disastrous end.

CHAPTER 2

PALEY AND KANT FIGHT BACK

HUME WAS TOO SUCCESSFUL. And he himself realized this. Right at the end of the *Dialogues,* having offered a series of withering arguments, the skeptic Philo (who scholars today think represented Hume himself) admits that there is something about the world which suggests more than pure chance. The argument from design (meaning both the argument to complexity and the argument to design taken together) does not do what its supporters think it does, according to Philo, but it does point to some unexplained mystery; the argument still has some force. It may be in bad shape, but it is not completely dead. If "the cause or causes of order in the universe probably bear some remote analogy to human intelligence," then "what can the most inquisitive, contemplative, and religious man do more than give a plain, philosophical assent to the proposition, as often as it occurs; and believe that the arguments on which it is established exceed the objections which lie against it?" (Hume 1779, 203–204).

The real problem, for Hume and others, seems to center around organisms. Final causes have not been strongly evoked by the physical world, nor deemed necessary for understanding it. Remember that Aristotle himself wrote on meteorology without resorting to final causes even once. But when it comes to organisms, function-talk not only occurs but seems essential. A rock may not have a purpose, but an eye does. Eyes and hands do not just happen for no reason. Plato's and Aristotle's worry about the atomists' inadequate account of how the world came to be is as valid after Hume as it was before. There has to be more. Yet what? What reply could be given, had to be given, to a critic such as Hume?

There were two major responses, one in England from the Archdeacon of Carlisle, William Paley, and the other in Germany from one of the most respected philosophers of modern times, Immanuel Kant. In their different ways, both men argued that there was still room for final causes and their inferences, despite Hume's skepticism. Let us take these responses in turn, remembering always that basically two questions are being raised. First, does the world have a level of complexity that seems to demand a special kind of explanation? Even if this is not true of the inorganic world, could it nevertheless be true of the world of organisms? Is there still a place for final-cause understanding? Second, assuming that such a level of complexity can be found, does it demand explanation in terms of a deity? Do final causes imply the existence of a god, and is this god the Christian God? In the terminology of Chapter 1, the argument to complexity and the argument to design.

The Anglican Compromise

William Paley (1743–1805), who at the beginning of the nineteenth century authored the standard work on the argument from design, did not spring from nowhere. He took his place in a British tradition going back several centuries to the Reformation and the Protestant break from Rome. At that time, Scotland turned Presbyterian, thanks to the proselytizing efforts of Calvin's lieutenant, John Knox. Ireland remained Catholic except for Northern Ireland, which became Protestant after the arrival of Scots Presbyterians. Wales eventually embraced variants of nonconformism—Baptism, Congregationalism, Methodism—all with roots in aspects of the Protestant Reformation. England itself produced a peculiar faith of its own: Anglicanism, embodied in the Church of England (also called the Anglican Church). The Episcopal Church was its American offshoot.

The Anglican Church is a strange hybrid. It is firmly Protestant in that it rejects the authority of the Pope and other defining Catholic attributes like a celibate clergy. Yet as England's "established" (state) church, it not only insists that the monarch be Anglican but also retains much of the ritual, hierarchy, ornamentation, and even the cathedrals of the Catholic Church. Though this "middle way" *(via media)* makes many Protestants uncomfortable to this day, it reflects the Church's origins, which were as much political as theological.

The broad outlines of this history are well known: Henry VIII (who ruled from 1509 to 1547) wanted to divorce his first wife, Catherine of Aragon, so that he could marry the pregnant Anne Boleyn in hopes of producing a legitimate son and heir. When the Pope in Rome refused to grant an annulment, Henry simply took his country and went home. Monasteries were destroyed and their property appropriated; objectors like Sir Thomas More lost their heads; the king got his new wife (who also in time was forced to place her own little neck on the executioner's block); and England was henceforth Protestant.

There is more to the story than this, of course. Breaking with Rome was no simple and straightforward matter. After Henry's death, his only son, the firmly Protestant Edward VI (child of third wife Jane Seymour, who died giving birth to Edward), became monarch until his early death. A Catholic revival ensued while Henry's oldest child, Mary—daughter of Catherine of Aragon, wife of the militant Catholic Philip II of Spain, and a devout Catholic from an early age—sat upon the throne (1553–1558). Mary persecuted Protestants, three hundred of whom she burned at the stake for heresy. Many others fled to safe havens on the continent until Bloody Mary was safely dead and buried. Finally came Henry's younger daughter, Elizabeth, with whom Anne Boleyn had been pregnant at the time of her marriage to Henry VIII. Under the long reign of the "Virgin Queen" (1558–1603), the Protestant Reformation was finally and firmly established in England.

Still, older families in parts of England like Norfolk in the east and Lancashire in the north remained faithful to Catholicism. These "recusants" were often aided by forces abroad, after the Pope allied with Spain, a Catholic country with major territorial designs. The old order posed an ever-present danger to Elizabeth. From the other side, tensions were generated by Protestant extremists, who had fallen under the influence of Calvin on the Continent during their exile under Mary I. Upon their return to England, they promoted a far more stern and rigorous faith than the comfortable path the home church was now happily walking. They wanted to move from the traditional, common-sense attitude that good works count above all to the sterner belief that faith alone saves the sinner from eternal damnation, and they wanted to put above the sovereign (who was head of the church) and her laws and ministers the ultimate authority of sacred scripture. These enthusiasts wanted to do away with such comforts as music in services and baptismal founts and to substitute endless, edifying sermons.

On the one side, therefore, Elizabeth faced the danger of the old Catholicism, along with the Jesuits and others who were doing their best to derail the Anglican Reformation. On the other side, she had to contend with the so-called Puritans, who were doing their bit to impede progress. What was needed was a safe middle passage between the Scylla of Catholic authority and the Charybdis of Puritan biblicalism. It was the genius of Elizabeth and her church to find just such a middle passage.

Socially and politically, Elizabeth anathematized Catholicism by thwarting the hostile invasion attempts of the devout and despised Spanish. Intellectually and theologically, the Elizabethan church downplayed Puritanism's faith in biblical scriptures and personal salvation by pointing out that the use of reason and evidence has biblical warrant. This is a major reason that the Church endorsed natural theology.

This strategy was embraced by the Oxford-trained cleric Richard Hooker, in his *Laws of Ecclesiastical Polity*. By turning to reason and evidence, one did not need to rely on Catholic authority and tradition. The truth was there, in nature, for all to see, if only they would exercise reason, observation, and good will. Nor, against the other extreme, did one need to rely on the unaided word of scripture. Indeed, Hooker said, it is an error to think that "the only law which God has appointed unto men" is the word of the Bible (*Works*, I.224).

Hooker was prepared to carry his promotional campaign for natural theology far beyond the bounds set by the theologians of the past, who considered it to be merely a support for revealed truths. For Hooker, natural theology was capable of going out in front alone—not just as a prop but as an essential player in the Christian drama. Scripture is without defect, but so is God's creation, properly understood. And, Hooker argued, it is part of our obligation to the Creator to study his creation, which can be understood only through sense and reason. "Nature and Scripture do serve in such full sort that they both jointly and not severally either of them be so complete that unto everlasting felicity we need not the knowledge of anything more than these two may easily furnish" (I.216).

Biological Contrivance

With its seemingly invincible argument in service to a pressing political need, Anglican natural theology was set for three centuries of triumphant dominance. It gave adherents the background and impetus to seek evidence of God's designing plan in the living world, and this led to an obsession with more scientific questions about organized complexity and final causes—questions which, if they were not asked in the physical sciences, were certainly felt to be essential for understanding organisms.

At first, as one might expect, there was considerable time lag and overlap between the older ways of doing things and the methods and metaphors of the newer sciences. Aristotle was not about to disappear overnight, nor was his way of thought inimical to all good new work. William Harvey—rightly celebrated as the person who recognized the heart as a pump which circulates blood around the body—had trained at Padua, a stronghold of Aristotelian medicine, and this influence stayed with him all of his life. From Aristotle, Harvey inherited the notion of the heart's primacy; by contrast, Galen had taught that the liver is the key organism in human physiology. Harvey's approach to the problem of circulation—with its attention to valves in the veins and the different chambers of the heart—was permeated with function-talk: with stress on what was best for, or of most value to, an organism and its parts.

As the new science started to take over, with its emphasis on analysis and empiricism, the appeal of final causes went into decline, especially in the physical sciences. But this decline was not at once complete and absolute. In France, for all that he argued against final causes as such, even Descartes was prepared to appeal to such teleological notions as the principle of least action, as was the number theorist Pierre de Fermat (1601–1665), whose name now symbolizes the principle: "Our demonstration rests on a single postulate: that nature operates by the simplest and most expedite ways and means" (Fermat 1891–1912, 3.173). Moreover, although this kind of value-impregnated thinking was on the decline as a tool of physical science, most physicists still expressed a general commitment to a God-backed and God-designed universe. Isaac Newton (1642–1727), discoverer of the universal law of gravitational attraction, was virtually intoxicated with the idea of God. Although he kept his speculations in this direction out of his scientific writings for the most part, in other contexts he was prepared to tie in his physics with an end-directed understanding that, he thought, pointed toward the deity. In a series of letters (1692–1693) to Richard Bentley, a younger contemporary and chaplain to the Bishop of Worcester, Newton wrote explicitly of the universe in terms that Plato would have understood and with which he would have sympathized. Everything we know about the universe, "the quantities of matter in the several bodies of the sun and planets, and the gravitating powers resulting from thence," not to mention "the several distances of the primary planets from the sun, and of the secondary ones from *Saturn, Jupiter,* and the Earth," and throwing in "the velocities, with which these planets could revolve about those quantities of matter in the central bodies," adds up to one conclusion and one conclusion only. The Cause behind everything "be not blind and fortuitous, but very well skilled in mechanicks and geometry" (*Opera Omnia,* 4.431–432).

When final-cause thinking seemed useful, British scientists were very ready to pick right up on it, particularly in the biological sciences. The point is that thinking of the world as a machine, and stressing the way in which it runs according to blind law, is all very well, but machines have machine makers, and the more intricate and complex the machine, the more pressing becomes the need to suppose a maker. An example constantly referred to was the elaborate clock on the cathedral at Strasbourg, built (between 1571 and 1574) by the Swiss mathematician Cunradus Dasypodius. It had little figures that revolved and danced on the hour. One cannot get away from intention there, and the same is true of organisms.

Although Robert Boyle conceded that intention was of little use in the physical sciences, he nevertheless argued strongly for its significance and propriety in the life sciences. Descartes in particular came in for severe criticism, and the human eye was offered as a counter-example to the "Cartesians." British naturalists felt a positive moral obligation to study nature and to work out its adaptations; in Boyle's words, "For there are some things in nature so curiously contrived, and so exquisitely fitted for certain operations and uses, that it seems little less than blindness in him, that acknowledges, with the Cartesians, a most wise Author of things, not to conclude, that, though they may have been designed for other (and perhaps higher) uses, yet they were designed for this use" (Boyle 1688, 397–398). From this complexity (what Boyle would call "contrivance") in the realm of science, he slides easily toward design in the realm of theology: "It is rational, from the manifest fitness of some things to cosmical or animal ends or uses, to infer, that they were framed or ordained in reference thereunto by an intelligent and designing agent" (p. 428).

An approach like this opens the door for the naturalists to give full reign to their imaginations—both in tying their discoveries in the living world to God's intentions and in using God's careful plan and execution as a device to understand the living world in ever greater detail. Prime among these was the clergyman-naturalist John Ray (1628–1705). In his *Wisdom of God, Manifested in the Words of Creation* (1691; 1709), he states the argument to complexity clearly and unambiguously. "Whatever is natural, beheld through [the microscope] appears exquisitely formed, and adorned with all imaginable Elegancy and Beauty. There are such inimitable gildings in the smallest Seeds of Plants, but especially in the parts of Animals, in the Lead or Eye of a small Fry; Such accuracy, Order and Symmetry in the frame of the most minute Creatures, a *Louse*, for example, or a *Mite*, as no man were able to conceive without seeming of them." Everything that we humans do and produce is just crude and amateurish compared with what we find in nature. Then comes the argument to design. The living world is likened to a product of design. A machine implies an architect or an engineer, and so likewise inasmuch as the world of life is machinelike, it too implies a being, as much above us as the world of life is above our artifacts and creations. "There is no greater, at least no more palpable and convincing argument of the Existence of a Deity, than the admirable Art and Wisdom that discovers itself in the Make and Constitution, the Order and Disposition, the Ends and uses of all the parts and members of this stately fabric of Heaven and Earth" (Ray 1709, 32–33).

A neat package, as the teleological way of thinking in biology is tied back into the proof of the divine: the classic argument from design. Moreover, this is design of absolutely top quality, and so the same must be said of the intelligence behind it. This points to a being worthy of worship, not to some ethereal, local spirit before which the heathen humble themselves.

The Nineteenth Century

After the social and religious upheavals of the seventeenth century—not just the Civil War but the deposition of the Catholic king, James II—natural theology settled into a prominent and vital role in the calming ideology of the established Anglican church. Ray's ideas were taken up and disseminated by his fellow clergyman William Derham, whose *Physico-Theology* had gone through thirteen editions by 1768. At the same time, the second half of the design argument was polished and extended by the moral theorist Joseph Butler (first Bishop of Bristol and then of Durham) in his *Analogy of Religion* (1736). He was less interested in making the argument as such—the fact of contrivance was taken as definitive—and more focused on the theological implications, particularly about how we can argue from pain, punishments, and rewards in this life to those we might expect in the hereafter. All in all therefore, both as science (argument to complexity) and as theology (argument to design), teleological thinking was vital.

As the eighteenth century drew to an end and the next began, the Scotsman David Hume might argue as he would, but in England natural history filled a more crucial social role than ever before. On the one side, there was the continuing threat of Catholicism—from France now more than Spain, as Napoleon began to test his strength on the Continent. Then on the other side, pressure came from within, from the disruptions caused by industrialization—including abandonment of the countryside in favor of jobs in the cities and the weakening influence of traditional institutions, especially the Church—and from Catholics in Ireland, whose agitation led to the Catholic Emancipation Act at the end of the 1820s, which allowed Catholics to hold public office and serve as members of Parliament.

Natural theology represented a calm, moderate, unifying, middle-of-the-road position. People were frightened of Hume's skepticism and of the rank atheism of the French *philosophes* (intellectuals); and with some justification, they saw science as being a fellow traveler with these threats. The more one stresses the universal laws of mechanics, then the more remote become things like miracles (Hume had criticized these no less than design) and the more a sensible Christianity seems under attack. This was no small worry for scientists, particularly those in England, for there education was entirely within the thrall of the church—particularly at the two universities, Oxford and Cambridge, where almost all of the dons had to be not just members of the church but ordained ministers. Scientists who wanted to support themselves by holding positions at these universities increasingly needed some counter-argument to show that the pursuit of science, far from threatening the true faith, strongly supported it. Natural theology, and design in particular, was the perfect answer. One could do one's science and at the same time claim that, through one's findings about the marvelous nature and workings of the empirical world, one was burnishing the most powerful argument there is for God's existence and perfect, all-powerful nature.

Paley, who had already established himself as a popular writer on matters of religion, offered the official response to Hume. First he defended revealed religion—the religion of faith—particularly in his counter-attack on Hume's critique of miracles. Then with the *Evidences of Christianity* out of the way, Paley turned to natural theology—the area of reason and argument. His *Natural Theology* of 1802 was the classic statement of the argument from design. The opening passage about the implications of finding a watch has become famous (Paley 1819, 1):

> In crossing a heath suppose I pitched my foot against a stone, and were asked how the stone came to be there, I might possibly answer, that for any thing I knew to the contrary it had lain there for ever; nor would it, perhaps, be very easy to show the absurdity of this answer. But supposing I had found a watch upon the ground, and it should inquired how the watch happened to be in that place, I should hardly think of the answer which I had before given, that for any thing I knew the watch might have always been there. Yet why should not this answer serve for the watch as well as for the stone; why is it not as admissible in the second case as in the first? For this reason, and for no other, namely, that when we come to inspect the watch, we perceive—what we could not discover in the stone—that its several parts are framed and put together for a purpose, e.g. that they are so formed and adjusted as to produce motion, and that motion so regulated as to point out the hour of the

> day; that if the different parts had been shaped different from what they are, or placed after any other manner or in any other order than that in which they are placed, either no motion at all would have been carried on in the machine, or none which would have answered the use that is now served by it.

A watch implies a watchmaker. Likewise, the adaptations of the living world imply an adaptation maker, a deity. The argument to design. You cannot argue otherwise without falling into absurdity. "This is atheism; for every indication of contrivance, every manifestation of design which existed in the watch, exists in the works of nature, with the difference on the side of nature of being greater and more, and that in a degree which exceeds all computation" (p. 14). This is the precise point that John Ray expressed a century before. Inasmuch as the workings, the contrivances, of nature exceed anything that we humans can produce, so then we must infer that the cause behind these contrivances is superior to us humans.

But now, of course, Paley realized that he must argue his case rather than simply state it. And once again the eye became a window into the soul of the machine. "I know no better method of introducing so large a subject, than that of comparing a single thing with a single thing: an eye, for example, with a telescope" (p. 14). Paley may not have been an original thinker, but his writing was vivid. The eye was a well-known example, but Paley knew how to make it serve another season. He knew how to capture his reader with a compelling description. He starts by explaining the workings of the eye itself in exquisite detail, spelling out its various parts. And then, having established the argument to adaptive complexity for the organic world, Paley takes a triumphant bow and simply infers the argument to design. "What plainer manifestation of design can there be than this difference? What could a mathematical instrument maker have done more to show his knowledge of his principle, his application of that knowledge, his suiting of his means to his end—I will not say to display the compass or excellence of his skill and art, for in these all comparison is indecorous, but to testify counsel, choice, consideration, purpose?" (p. 15).

Was Paley still with Hooker in thinking of the argument from design as a definitive argument for God's existence and wonderful nature, or had he slipped back to the thinking of earlier times? Did he regard it as merely an aid to faith, serving more as a supplement to a conviction already achieved on nonrational grounds than as a primary argument capable of convincing nonbelievers? Paley stood firmly on the fence. At one level, Paley was unambiguously in the English Protestant position, taking the argument from complexity as definitive, as proving God beyond doubt. It is an alternative to faith, not a handmaiden. "Were there no example in the world of contrivance except that of the eye, it would be alone sufficient to support the conclusion which we draw from it, as to the necessity of an intelligent Creator. It could never be got rid of, because it could not be accounted for by any other supposition which did not contradict all the principles which we possess of knowledge—the principles according to which things do, as often as they can be brought to the test of experience, turn out to be true or false" (p. 59).

Yet, at another level, Paley saw the argument from design as having an ongoing function. It could continue to enrich and inform the life of the believer. When once we see the hand of the Creator in his creation, "the world henceforth becomes a temple, and life itself one continued act of adoration. The change is no less than this: that whereas God was formerly seldom in our thoughts, we can now scarcely look upon any thing without perceiving its relation to him" (Paley 1802, 420–421). Paley set the tone, provided a pattern, and offered a religious justification for pursuing a life of inquiry and scientific endeavor in the nineteenth century.

There were strong social reasons why his arguments were accepted gratefully, but Paley was also relying on the matter that had troubled Hume himself: namely, that simply being critical leaves an unacceptable gap. Paley was not really offering a simple inductive argument: the world is artifact-like, ergo it very probably has a designer. This lays itself open to the Humean charge that the world is not really artifact-like at all. Rather, Paley offers what is known as an "abduction" or "inference to the best explanation": the world (the organic world especially) has features that need explanation, and the only viable explanation is that of design. However much you may (truly) say that the world is not artifact-like, if it has the key features that need explanation, then you are home free. To Paley's intellectual credit, he did raise various other possibilities, for instance that adaptation was produced by chance. But he thought them ineffective and dismissed them. "What does chance ever do for us? In the human body, for instance, chance, *i.e.* the operation of causes without design, may produce a wen, a wart, a mole, a pimple, but never an eye" (Paley 1802, 49).

Moreover, Paley was not getting away from any kind of analogical argument. He had to make the case that the eye did have design-like features that need explanation. But, this done, he could waltz home: "We conclude that the works of nature proceed from intelligence and design; because, in the properties of relation to a purpose, subserviency to a use, they resemble what intelligence and design are constantly producing, and what nothing except intelligence and design ever produce at all" (p. 325). If design remains the unique explanation that can do the job, then at one level all of the counter-arguments put forth by Hume fall away. As Sherlock Holmes, speaking to his friend Dr. Watson, put it so well: "How often have I said to you that when you have eliminated the impossible, whatever remains, *however improbable,* must be the truth."

Obviously, looking at things with hindsight, Paley's was something of a stopgap position. Because he and everyone else was ignorant of an adequate alternative to direct divine intervention, this does not mean that Hume's arguments were bad. It was just that back then they could not have the definitive force that many today think they deserve. As soon as an alternative appears—as soon as design is not the unique explanation—then the Humean critique could come into play and help us to decide about how to account for contrivance. But in the alternative's absence, the Holmesian point applies.

The fire that Paley banked continued to smolder and smoke and then burst into bright flame some decades later, thanks to an 1829 bequest by the eighth Earl of Bridgewater. At his death this Anglican clergyman left £8,000 to sponsor the writing of a series of works "On the power wisdom and goodness of God as manifested in the creation." Eminent men of science were enrolled to do their duty, and with the subsequent appearance in the 1830s of the eight Bridgewater treatises, British natural theology reached its apotheosis—or, given the way that some authors went right over the top and down the other side, its nadir (Gillespie 1950).

A typical essay was the offering on mineralogy and geology by the Reverend William Buckland, professor of geology at Oxford. Buckland was famous for his party trick of leading large, mixed-sex audiences into subterranean caves, where he liked to deliver popular lectures on fossils dressed in top hat and frock coat and to conclude them with a singing of the national anthem. In his Bridgewater treatise, Buckland blended God, King, and country in a heady brew. Drawing the reader's attention to the world's distributions of coal, ores, and other minerals, he concluded that this showed not only the designing nature of a wise Providence but the especially favored status of a small island off the coast of mainland Europe. "We need no further evidence to shew that the presence of coal is, in a especial degree, the foundation of increasing population, riches, and power, and of improvement in almost every Art which administers to the necessities and comforts of Mankind." It took much time and forethought to lay down those strata, but their very existence, lying there for "the future uses of Man, formed part of the design, with which they were, ages ago, disposed in a manner so admirably adapted to the benefit of the Human Race" (Buckland 1836, 1.535–538). It helped, of course, that God was an Englishman. The location of vital minerals "expresses the most clear design of Providence to make the inhabitants of the British Isles, by means of this gift, the most powerful and the richest nation on earth" (Gordon 1894, 82).

As Buckland was singing hymns to Britain's favored status in God's eternal plan, a young fellow countryman, Charles Darwin, was working secretly on a theory that would make all such teleological boasting look very thin indeed. But even before Darwin published, some of Buckland's fellow theologians and scientists were worrying publicly that the confidence of Anglicans in their tradition of natural theology was misplaced or at least overvalued. Even when they ignored Hume's critique, these contemporaries were beginning to realize that this venerable tradition was looking old-fashioned and problematic.

However, this is to rush ahead of our story, for we must first go back to the eighteenth century and look at a reaction to Hume's critique that was very different from Paley's apparently confident restatement of the traditional position. This opened the way to a different approach to organisms in terms of their ends and functions, although in the long run, it, like the Anglican tradition, could not halt the progress of Darwinism. And in a sense it helped make the Darwinian approach more secure and successful.

Immanuel Kant

The German philosopher Immanuel Kant (1724–1804) always acknowledged how David Hume had "awakened him from his dogmatic slumbers," meaning that Hume's critical thinking had forced him to search for alternative answers to the skeptical conclusions of the Scotsman. Kant's "Copernican revolution" in philosophy centered on the argument that our knowledge is not simply read off from experience but is in an important sense formed and structured by our own minds. Hume had boiled everything down to psychology: there is no definite truth, and all we can do is live according to our nature, which fills us with illusions about the objectivity of knowledge and morality, and that is not only just the best we can do but quite sufficient for our daily needs. Kant wanted more than this, and although he accepted completely the Humean position that we can never return to the certainty of the old days, he thought that some kind of necessity and objectivity were possible. This comes through the fact that although we ourselves interpret nature according to certain constraints, this act of interpretation is in some sense the necessary condition for any possible rational thought by intelligent beings.

Thus Kant went beyond Hume. While agreeing that it is we ourselves who put certainty and truth and objectivity into nature, rather than reading them from nature, he claimed that the way in which we interpret is no mere contingent fact about our human nature. The way itself is privileged: only in this fashion is it possible to think and act at all—at least, to think and act as rational beings. To think causally, for instance, is a requirement for any kind of coherent sensible thought, and to behave morally is simply something that we have to do if we are to live successfully as social beings. If we do not think both causally and morally, then we simply cannot function. A group of totally immoral people would not be able to live together and survive, for relationships would break down entirely. For Kant, causality is not a logical necessity; just as there is no analytic contradiction in thinking that the law of gravity may fail, so it is not illogical to think that causality may fail. But causality is more than mere sense experience; we interpret and structure the world from within ourselves. Hence, causal claims are in some sense not just given in experience, *a posteriori,* but come from thought, *a priori.*

Kant did not think that a God, proven in traditional ways, could rescue matters. A man of deep religious sensitivity—he was raised by devout pietistic parents—Kant nevertheless conceded that Hume and others had destroyed the validity of the traditional arguments for God's existence, including the argument from design. In their place, Kant posited a God whose existence must be presupposed if we are to make sense of much that we experience, and that God is especially needed for the conviction that morality is something of worth and validity. "I have therefore found it necessary to deny *knowledge,* in order to make room for *faith.* The dogmatism of metaphysics, that is, the preconception that it is possible to make headway in metaphysics without a previous criticism of pure reason, is the source of all that unbelief, always very dogmatic, which wars against morality" (Kant 1781, 29).

One of the most important and controversial aspects of Kant's philosophy—something which has given rise to two hundred years of discussion and revision—is a distinction he drew between the world of appearance (the phenomenal world) and the world of reality (the noumenal world). It is only the former we can experience, but we must suppose the latter (which is the world of God) in order to resist the skepticism of Hume. This second world—what Kant calls the world of "things in themselves"—plays a crucial role in Kant's solution to such traditional philosophical problems as free will and determinism. Much impressed by the Scientific Revolution and in particular the achievements of Newton, Kant was convinced that the whole physical (phenomenal) world, including us humans, is subject to the invariable laws of nature. We are all therefore part of the determined causal network. Yet we are moral beings and so in some sense must be free. This moral freedom is something that we experience in the noumenal world. In this argument Kant offers the ultimate basis of morality and the ability to choose and to exercise our freedom.

Kant was keenly interested in organisms and the sciences devoted to their study. Being acquainted with the work of Georges Leclerc, Comte de Buffon, the famous naturalist of mid-eighteenth-century France, he knew of theories of generation and was aware of the evolutionary speculations of some of his contemporaries. He himself dabbled in such speculations in cosmology, as when he helped to formulate the nebular hypothesis—the idea that the universe came into being naturally from large masses of gas, called nebulae, in accordance with Newtonian processes.

Kant noted that biologists of his day regularly referred to final causes, even though such simple appeals to design were no longer acceptable. This motivated him to offer a definition of the kind of case in nature where it is appropriate to use final-cause thinking. He tried to characterize what it is that we find special about the organic world, and in special need of explanation. He focused on what we, along with today's evolutionists, are identifying with the term adaptive complexity.

The complexity is all a question, apparently, of something's being both cause and effect of itself. Kant (1790, 18) gave an example to show what he had in mind. A tree produces another tree. But as it does so, it produces another specimen of the same genus as itself. Hence the genus is both cause and effect, producing itself and being produced by itself. And this is true at the individual level. "The plant first prepares the matter that it assimilates and bestows upon it a specifically distinctive quality which the mechanism of nature outside it cannot supply, and it develops itself by means of a material which, in its composite character, is its own product." Finally, "the preservation of one part is reciprocally dependent on the preservation of the other parts." The tree produces the leaves, but at the same time the tree is dependent on the leaves. Defoliate the tree and you put it at risk of death. Cause is a two-way street.

This is not to say that Kant was now arguing for reverse causation. One may understand the cause in terms of the effects, but from a temporal perspective what one has is cause bringing on effect, which is now cause for an effect, which possibly will in turn be a cause for an effect, and so forth. This connects with what we have been saying about complexity, because it is in and only in complex systems that we have sufficient causal subtlety or sophistication that something can bring about effects which seem to be needed for the cause itself in the first place—the tree bringing about the leaves that the tree needs.

Kant was uncomfortable with final-cause talk because it does seem to imply design, and this is simply not acceptable in science. We are only allowed to talk in terms of material or mechanical (that is, efficient) causes. "Hence if we supplement natural science by introducing the conception of God into its context for the purpose of rendering the finality of nature explicable, and if, having done so, we turn round and use this finality for the purpose of proving that there is a God, then both natural science and theology are deprived of all intrinsic substantiality" (p. 31). But Kant recognized that we simply cannot do without final-cause thinking. In biology, teleology is absolutely essential. We need the maxim: "An organized natural product is one in which every part is reciprocally both end and means." In short, we simply cannot do biology without assuming final cause. "It is common knowledge that scientists who dissect plants and animals, seeking to investigate their structure and to see into the reasons why and the end for which they are provided with such and such parts, why the parts have such and such a position and interconnexion, and why the internal form is precisely what it is, adopt the above maxim as absolutely necessary."

In Kant's view, scientists cannot do biology in any other way. Teleological thinking is not a luxury; it is a necessity. Life scientists are as bound to teleology as they are to physics. "For just as the abandonment of the latter would leave them without any experience at all, so the abandonment of the former would leave them with no clue to assist their observation of a type of natural things that have once come to be thought under the conception of physical ends" (p. 25).

So how are we to solve the problem (which Kant called an *antimony*) of needing to use final-cause talk and yet recognizing that only material-cause talk is acceptable in any science that claims to be speaking of objective reality? Here the Kantian metaphysics comes into play—phenomenally we can see no design in nature, but noumenally it is possible that there is design. God may be standing behind everything, but this is for things in themselves and not for the phenomenal world as we know it. We may (must) suppose God, but we cannot prove it. "All that is permissible for us men is the narrow formula: We cannot conceive or render intelligible to ourselves the finality that must be introduced as the basis even of our knowledge of the intrinsic possibility of many natural things, except by representing it, and, in general, the world, as the product of an intelligent cause—in short, of a God" (p. 53).

For Kant, then, final causes are not a condition of rational thinking in the way that mechanical physics is. We cannot think of the world except as causally. We can certainly look at organisms without thinking of final causes, but as soon as we start to study them, to understand them, final-cause thinking comes into play—*has* to come into play. Like Aristotle, Kant thought that the organic world must be understood teleologically. For Aristotle, however, final causes are out there, part of reality; they have an ontological status of their own. For Kant, final causes are the very lens through which we observe and study the world. They are our doing—similar to things like causality in that we impute them to the world but less strong than causality because we can think without them even though we cannot work without them. They are regulative. "Strictly speaking, we do not observe the *ends* in nature as designed. We only read this conception *into* the facts as a guide to judgement in its reflection upon the products of nature. Hence these ends are not given to us by the Object" (p. 53).

As far as the argument to complexity is concerned, Kant viewed objects in the living world as being both cause and effect. As far as the argument to design is concerned, we suppose God but we do not prove his existence. We must think in terms of ends, but it is we who put ends into nature; we do not find them there as existing entities. Paley was willing to keep the divine consciousness in or around his picture of the world. And this world has value because it is the creation of a good God—in fact, Paley (1802) went into some detail to show that Hume's worries about evil and pain are not as bad as one might think. The values have to be the best possible on some absolute scale, and every unpleasant cloud has a silver lining. ("A man resting from a fit of the stone or gout is, for the time, in possession of feelings which undisturbed health cannot impart," p. 321). With Kant, we must see the world as if it were produced with intention. Although Kant had faith that there is value in the world, as a reasoning, thinking being he found no value there to be read from the world. It is a value that we must impart to the world—or rather, values that we must see in the workings of things toward ends.

CHAPTER 3

SOWING THE SEEDS OF EVOLUTION

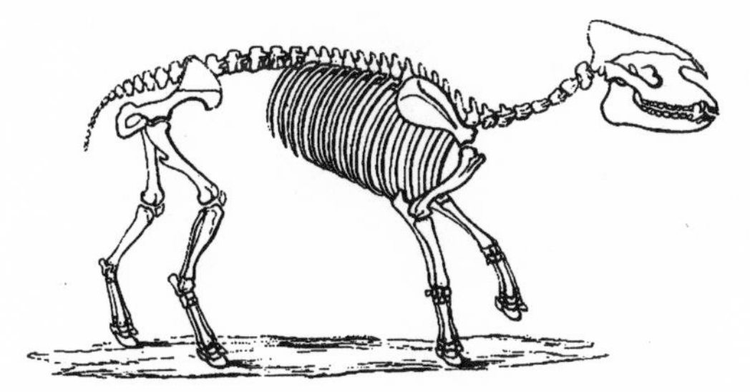

ALTHOUGH THE ATOMISTS believed in natural origins for organisms, they did not believe in a gradual naturalistic development from just a few primitive forms. So while they were not evolutionists, perhaps one can describe them as proto-evolutionists—a term that distinguishes them from Plato and Aristotle, who had no place at all for such developmental thinking. The coming of Christianity with its Genesis story of creation reinforced a static view of origins. But the Old Testament also gave a historical dimension that had not existed hitherto, and this was significant in pressing the need for some account of the beginnings of things.

Not that one should consider the earliest Christian thinkers, the Church Fathers, to be biblical literalists in the sense that modern American creationists are. But in the absence of a persuasive counter-argument, neither Saint Augustine nor his successors saw reason to doubt that God had miraculously created the heavens and the earth along with all organisms, including us humans. Still, Augustine in particular allowed room for metaphorical interpretation. Since he thought that God stands outside time and that hence the idea, the act, and the finished product of creation are to God as one, Augustine believed that life developed from seeds of potentiality already in place from the first moment. Catholics today find in this interpretation a theology congenial to evolutionism, but we should not pretend that Augustine himself was thinking scientifically in advance of his time.

Holes in the dike of miraculous creation started to appear after the Scientific Revolution. Partly, this was because of movements in the physical sciences, as we have seen, where even Kant began to speculate about the evolution of stars and planets. And partly, it was because of recent empirical discoveries: fossils were being unearthed and identified as the remains of long-dead organisms; geographical variations in the distribution of plants and animals were being revealed; and fabulous new beasts—including those semi-humanlike pongos and jockos, today more prosaically known as monkeys and apes—were being brought back to Europe. But mainly, disbelief in miraculous creation was breaking down because of socioeconomic ideas and philosophies that were beginning to undermine traditional Christianity.

Above all, the crucial factor was a growing enthusiasm in the eighteenth century for the philosophy of progress—the belief that we humans through unaided effort can improve our knowledge and, through science and technology, our lot in life. This kind of end-directed plan is a human plan, not a divine plan. Thus, the notion that progress could be made at a purely secular level was explicitly opposed to the Christian notion of Providence, which supposed that nothing happens without God's grace and that without his help, our greatest efforts count for naught. Progress moves ever onward and upward through human-driven ingenuity, work, and commitment. Not surprisingly, it was a very easy move from notions of progress in the human social and intellectual world to the idea of progress in the living world; and once this was seen as evolution, it became very easy to circle back to the notion that evolution in the natural world confirms progress in the human world.

Erasmus Darwin Embraces Evolutionary Progress

Erasmus Darwin, grandfather of Charles, was the quintessential British progressionist of this period. A physician in the British Midlands in the second half of the eighteenth century, he was carried along by the first wave of the Industrial Revolution, as enterprising engineers put coal and steam to work running machines that produced finished goods at a far faster rate than anyone could ever make by hand. On a trip to observe the construction of a new canal tunnel—"I have lately travel'd two days journey into the bowels of the earth, with three most able philosophers, and have seen the Goddess of Minerals naked, as she lay in her inmost bowers" (letter to Josiah Wedgwood, July 2, 1767)—Erasmus Darwin was much influenced by the fossils he saw along the walls of the newly cut earth. With this evidence at hand, he moved readily from a belief in social and industrial progress to belief in progress within the organic world. Evolution from the primitive to the complex, from "monad to man," as the popular phrasing had it, from monarch (butterfly) to monarch (king), as he himself once put it. "Would it be too bold to imagine, that all warm-blooded animals have arisen from one living filament, which THE GREAT FIRST CAUSE embued with animality, with the power of acquiring new parts, attended with new propensities, directed by irritations, sensations, volitions, and associations; and thus possessing the faculty of continuing to improve by its own inherent activity, and of delivering down those improvements by generation to its posterity, world without end?" (Darwin 1801, 2.240).

Actually, when it came down to the causes of organic evolution, Erasmus Darwin tended to be a bit fuzzy. Coming from the agricultural part of England, he was familiar with the breeding of horses, dogs, and cattle, and he was sure that the "great changes introduced into various animals by artificial and accidental cultivation" were significant (Darwin 1801, 2.233). That is, the inheritance of acquired characteristics was clearly a crucial factor in Erasmus Darwin's view of evolution. Also, he wrote of changes that are brought on first artificially by the natural environment or by human action but later got ingrained in heredity, as it were. He gave as an example the docking of dogs' tails, which (he claimed) ends eventually in animals born naturally with little or no tail.

The elder Darwin was always happy to throw in a few anecdotes, including some that would not be amiss on the *Jerry Springer Show.* Somewhat puzzling is his recipe for producing children of the desired sex and of great beauty. Not only does one need artificial penises worn in the hair, but "the fine extremities of the seminal glands" need to imitate "the actions of the organs of sense either of sight or touch." Unfortunately, for the curious, "the manner of accomplishing this cannot be unfolded with sufficient delicacy for the public eye," even though they "may be worth the attention of those, who are seriously interested in the procreation of a male or female child" (2.270–271).

Although Erasmus Darwin occasionally went over the top, in many respects as an evolutionist he was entirely typical. A year or two later, in France, Jean Baptiste de Lamarck promoted an evolutionary world picture that bore remarkable similarities to that of the elder Darwin. His *Philosophie Zoologique* of 1809 appealed to many of the same causal mechanisms, including the inheritance of acquired characteristics. Today—somewhat paradoxically—this is known as Lamarckism, although it was a prominent part of Erasmus Darwin's slightly earlier speculations and indeed is an old idea going back at least to biblical times.

Although Erasmus Darwin was no Christian nor an enthusiast for the theology of the Anglican Church, like many people of his age he believed in an Unmoved Mover—a God who had set things in motion and then stepped back from his handiwork. Darwin was a deist, meaning that he saw God's glory and power lying in the ability to do everything through unbroken law. There is no need of miracle or divine intervention. Here, Darwin stood opposed to the theist—traditionally the Christian, Jew, or Muslim—who emphasizes the significance of God's miraculous intervention in the creation. In the language of the present day, Darwin's god had preprogrammed the world to run without his ever having to intervene. Hence, for Darwin, far from evolution being a stain on his religious vision, it was its ultimate proof—its apotheosis. Here was a god truly worthy of worship.. In Darwin's words: "What a magnificent idea of the infinite power of THE GREAT ARCHITECT! THE CAUSE OF CAUSES! PARENT OF PARENTS! ENS ENTIUM!" (1801, 2.247). The proof of God's power came simply in the very fact that laws operate and produce effects. Darwin was not looking for proof of a miraculous, hands-on creation in the exquisite workings of the eye or the hand.

This is not to say that Darwin was against a God-driven purposeful perspective on the world of life. Indeed, one might say that, from a historical perspective, no one was more divinely teleological than Erasmus Darwin (except maybe Lamarck). His whole worldview was directed toward the end of humankind. Evolution, for him, was not a slow meander going nowhere. It was deeply directional, culminating in our own species. The coming of *Homo sapiens,* planned by the Creator, made sense of all that had gone before, and in its absence the world would have been inexplicable, in Darwin's view. In this respect, Erasmus Darwin was no less end-directed than the Christian who looks to the Day of Judgment, when the dead will rise and stand before their Maker. And no less value-impregnated, because the nature and fate of humankind were of extreme interest and importance to him.

Erasmus Darwin (along with Lamarck) was certainly not totally insensitive to adaptation and to the natural theologians' understanding of it in an end-directed way. A strengthened blacksmith's arms, inherited from his blacksmith father, has clear adaptive value—it is a "contrivance." Darwin's major prose work, *Zoonomia,* contains a remarkable sketch of what his grandson Charles was to label *sexual selection,* and it is rife with end-directed language and understanding. Erasmus Darwin wrote of male animals having "weapons" to fight other males, and how "the horns of the stag are sharp to offend his adversary, but are branched for the purpose of parrying or receiving the thrusts of horns similar to his own, and have therefore been formed for the purpose of combatting other stags for the exclusive possession of the females" (Darwin 1801, 2.237). He concluded: "The final cause of this contest amongst the males seems to be, that the strongest and most active animal should propagate the species, which should thence become improved" (2.238).

But adaptation was far from Erasmus Darwin's main focus, and in support of his evolutionism he made explicit reference to a category of phenomena (already noted by Aristotle) that throws doubt on the ubiquity of adaptation. These are the strong similarities in form between the parts of very different animals and plants, such as the forelimbs of vertebrates. The arms of humans, the wings of birds, the front legs of horses, the flippers of whales, the paws of moles—all are used for very different purposes but bear similarities in their structures. There seems to be no final cause at work here. Such isomorphisms—what we now call *homologies*—do not exist for the sake of some particular end.

So much for the argument to adaptive complexity. What about the argument to design and a divine designer? As a physician and a progressionist, as well as an evolutionist, Erasmus Darwin was not very much interested in this argument—the argument, that is, from complexity to a creative intelligence. He would have accepted it, but it would have been part of his overall commitment to a lawful universe. It would not have been the center of his world picture, as it was for Paley. No doubt Erasmus Darwin would have agreed with the updating of Robert Boyle's position by the mid-nineteenth-century theologian and mathematician

Baden Powell (father of Robert Baden-Powell, the founder of the Boy Scouts), who would argue that a world created through law is far superior to a world created by hand, that is, by miracle. "Precisely in proportion as a fabric manufactured by machinery affords a higher proof of intellect than one produced by hand; so a world evolved by a long train of orderly disposed physical causes is a higher proof of Supreme intelligence than one in whose structure we can trace no indications of such progressive action" (Powell 1855, 272).

Of course, a cynic might point out that Powell had to argue this way—to affirm the ultimate meaning of final cause, albeit set one step back—for Powell was an Anglican priest. For Erasmus Darwin, there was no such social pressure. As a deist but not a Christian, he would have given the Creator some ultimate designing power, but the pressures to be explicit were not strong. As an evolutionist, he played down the kind of final causes that Aristotle, Kant, and Paley espoused. This pattern was repeated by later evolutionists, from Lamarck to Robert Chambers, a Scottish businessman who anonymously authored the *Vestiges of the Natural History of Creation* (1844). This work caused a major evolutionary scandal in early Victorian Britain and set the background against which Darwin's *Origin* was to be judged. The fact is that these evolutionists were not obsessed with contrivance—that is, complexity. What they did care about was progress, human progress here on earth, and the argument to complexity had to take a back seat to that.

Immanuel Kant Rejects Evolutionary Thinking

But, as we know, others were indeed obsessed with contrivance, and what we must ask now is if this made a difference in their thinking about evolution. The answer is that it most certainly did, and this in fact was one of the major reasons why evolutionism was found unacceptable by thinkers with more respectable credentials than Erasmus Darwin's—thinkers, that is, whose enthusiasm for progress was not so wild and unrestrained that it could turn them into evolutionists, whatever the philosophical objections. Kant was a paradigmatic opponent of evolution precisely because of his commitment to adaptation and final causes. Given his interest in the nebular hypothesis, one might think that Kant would have been at least empathetic to evolutionary speculations. However, far from seeing evolution as in some sense the answer to the nature of organisms, and perhaps throwing light on final causes, he saw final causes as a barrier to the possibility of evolution.

Kant did not think that the idea of organic evolution was silly. Indeed, like Erasmus Darwin (whose specific thinking he seems not to have known), Kant thought that homologies between the parts of different organisms point in this direction. "This analogy of forms, which in all their differences seem to be produced in accordance with a common type, strengthens the suspicion that they have an actual kinship due to descent from a common parent." Kant even went on to spell things out, speaking of our ability to "trace in the gradual approximation of one animal species to another, from that in which the principle of ends seems best authenticated, namely from man, back to the polyp, and from this back even to mosses and lichens, and finally to the least perceivable stage of nature" (Kant 1928, 78–79). For Kant, there was nothing *a priori* self-contradictory about the idea of organic evolution, and "there are probably few even among the most acute scientists to whose minds it has not sometimes occurred." It is certainly logically possible that animals move from the water to the marshes and thence to the land, changing and adapting as they go along. Such a notion of evolution is not like a round square, which never could exist, even in principle. Nevertheless, Kant asserted, the facts of nature go against evolution. "Experience offers no example of it. On the contrary, as far as experience goes, all generation . . . brings forth a product which in its very organization is of like kind with that which produced it" (Kant 1928, 79n–80n).

But then Kant made it clear that the trouble with evolution goes deeper than that. Evolution may not be a logical impossibility, but it does go against the final-cause thinking we use in biology. Organisms are organized in the way they are for a reason, and moving across from one species to another would disrupt this organization in a way fatal to the intermediaries. "For in the complete inner finality of an organized being, the generation of its like is intimately associated with the condition that nothing shall be taken up into the generative force which does not also belong, in such a system of ends, to one of its undeveloped native capacities." Breaking from this inner finality would be disruptive for survival:

"The principle of teleology, that nothing in an organized being which is preserved in the propagation of the species should be estimated as devoid of finality, would be made very unreliable and could only hold good for the parent stock, to which our knowledge does not go back" (p. 80n). For Kant, it came down to choosing between evolution or final cause, and final cause won. The contrivance complexity of nature demands that we think in terms of ends appropriate to a designing intelligence. Evolution, by contrast, is a blind-law explanation, and no blind-law explanation can yield phenomena requiring understanding in terms of ends appropriate to an intelligence. Hence, evolution cannot in principle explain contrivance, and so must be false.

Cuvier Explains the Conditions of Existence

The most influential scientist in early nineteenth-century France was born in one of the border provinces between France and Germany and was educated in Germany. There, Georges Cuvier felt the full force of Kant's philosophy and, in part due to that influence, eventually became the greatest opponent that evolution ever faced.

In order to grapple with Cuvier's thinking, we must start with the fact that he was for many years one of the two all-influential permanent secretaries of the French Academy of Science—as much a bureaucrat as a professional scientist. In that role, Cuvier was determined to upgrade biological science to the level of the best science of his day, namely physics and chemistry. To this end, he sought tools or methods that would enable biologists to classify material and subordinate it to universal laws, opening the way for a truly unified world picture. He saw much of the life science of his day as being inexcusably flabby (which it was), and he was determined to recondition it so that it exhibited the epistemic merits of the best sciences, in particular, their ability to predict outcomes.

In some respects, given Cuvier's lofty aims for biology, Kant was an unlikely influence. The German thinker not only opposed evolution on philosophical grounds but probably doubted biology's potential for ever being a full-blown science. He thought it could, at best, be a group of generalities rather than a formal system of laws. But people have ways of using others' ideas to their own ends, and so it was with Cuvier and

Kant. The German's influence on the Frenchman was at least two-pronged: first, in the conviction that the key to understanding the living world is through recognizing its organized, end-directed complexity (living things are not just randomly put together); second—and this ties in, however paradoxically, with Cuvier's desire to produce a professional science—with the conviction that whatever one's personal religious beliefs, they can have no place in one's formal science. The approach of an English naturalist-parson like John Ray, for instance, was simply not acceptable to Kant or Cuvier. A science must stand alone.

With this mind-set, Cuvier responded very positively to the writings of Aristotle, whose works he had studied in some detail during the dangerous years of the French Revolution, which Cuvier had spent tucked away in Normandy. He saw in the internal teleology of the Greek philosopher the possibility of a mature, functioning biology; he wrote to a friend that Aristotle's writing on animals (the *History of Animals* and the *Parts of Animals*) was the "first essay in scientific natural history" (Cuvier 1858, 71). For the German-trained Cuvier, as for Aristotle, the key to understanding the organism lies in the fact that it is not simply subject to the physical laws of nature but that it is organized, with its parts directed to the end of the functioning whole—each individual feature playing its role in the overall, end-directed scheme of things. Cuvier referred to this organization as the "conditions of existence" (1817, 1.6):

> Natural history nevertheless has a rational principle that is exclusive to it and which it employs with great advantage on many occasions; it is the *conditions of existence* or, popularly, *final causes.* As nothing may exist which does not include the conditions which made its existence possible, the different parts of each creature must be coordinated in such a way as to make possible the whole organism, not only in itself but in its relationship to those which surround it, and the analysis of these conditions often leads to general laws as well founded as those of calculation or experiment.

Notice the reference to general laws, to calculations, and to experiment—all things that one finds in the better class of science. But how is one to translate this philosophy into a practical working science? Here we move to Cuvier's field of scientific interest and expertise, animal morphology (anatomy). He made detailed studies of animal after animal, thinking that the conditions of existence yielded a working guide for the investigator. This corollary, as we might call it, was referred to by Cuvier as the "correlation of parts." He argued that in order to be an integrated functioning being, every part of an organism had to be slotted in harmoniously with every other part. "It is in this mutual dependence of the functions and the aid which they reciprocally lend one another that are founded the laws which determine the relations of their organs and which possess a necessity equal to that of metaphysical or mathematical laws" (pp. 67–68). And then, linking this back to the conditions of existence (of which the correlation of parts is really just a physical manifestation), he wrote: "It is evident that the seemly harmony between organs which interact is a necessary condition of existence of the creature to which they belong and that if one of these functions were modified in a manner incompatible with the modifications of the others the creature could no longer continue to exist" (pp. 67–68).

The point was that the correlation of parts supposedly gives the anatomist a tool to make predictions. Pretend, for instance, you have only the tooth of an animal. From the design of that tooth you can tell that it is for tearing or slicing meat rather than for chewing vegetable matter. From that, you work outward. If the tooth's owner is a carnivore, you can be sure that it will not have hooves like a deer or a stomach like a cow or the armor of a tortoise or any of the other attributes that one associates with vegetarian prey. Rather, it will display claws, along with agility, intelligence, and so forth. Thus you can infer what the whole animal must look like. Cuvier felt that this line of argumentation showed its worth again and again, for on several occasions, having been given but a fossil fragment of an unknown beast, he was able to infer the whole form—a prediction triumphantly confirmed when later the whole animal was discovered.

The conditions of existence yield a second corollary, the "subordination of characters." It was through this that Cuvier thought he was able to bring order to the animal world, dividing it into four basic groups or embranchments (Cuvier 1817, 1.10):

> The parts of an animal possessing a mutual fitness, there are some traits of them which exclude others and there are some which require others;

> when we know such and such traits of an animal we may calculate those which are coexistent with them and those which are incompatible; the parts, properties, or consistent traits which have the greatest number of these incompatible or coexistent relations with other animals, in other words, which exercise the most marked influence on the creature, we call important traits, dominant traits; the others are the subordinate traits, and there are thus different degrees of them.

Suppose you have discovered a backbone. From it, you know that many features to be found in the animal world are impossible for this particular animal. It must now have characteristics and only those characteristics that are part and parcel of being a vertebrate. Suppose what you have is the backbone of a whale. It is not possible for this animal to have the limbs of a land predator like a tiger, nor the teeth of such an animal, nor its stomach or its brain, or many other things. Once you have gone the route of sea mammal, then for many characteristics you are constrained to one outlet only. This gives rise to the proper way to classify—being ever more restricted in the choices open to the production of a functioning organism.

And so we get the four basic groups (embranchments) of animals: the vertebrates (having a backbone), the mollusks (having a distinct brain and no central nerve chord), the articulates (having a small brain and two distinct central ventral nerve threads), and the radiates (having radial symmetry). "I have found that there exist four principal forms, four general plans, upon which all of the animals seem to have been modeled and whose lesser division, no matter what names naturalists have dignified them with, are only modifications superficially founded on development or on the addition of certain parts, but which in no way change the essence of the plan" (p. 92).

This, in a nutshell, is Cuvier's brilliant program for understanding and classifying the animal world. Aristotelean without a doubt, but also very much in the spirit of Kant, whose philosophy Cuvier would have found more attractive than that of the Greek. Not just a practicing Christian but a Protestant (the dominant religion of his home province), Cuvier would have responded empathetically to Kant's attempt to find a place for God—the God who, in some sense, would have been responsible for the design one finds in nature. However, although Cuvier's is a Christian, caring God (unlike the deity of Aristotle), his God also has to be taken on faith (as does Kant's) and does not play a direct role in our scientific understanding. In Cuvier's approach to final causes, the order and understanding are imposed by us on the world, rather than being a constituent of existence.

Whatever the difference between Kant and Cuvier on the possibility of a mature biology, Cuvier was right in line with Kant's denial that final causes and evolution can be held consistently by one and the same person. Like the philosopher, Cuvier was happy at one level to appeal to the empirical evidence. Napoleon had gone campaigning in Egypt and—taking his savants with him—returned to France with mummified humans and animals. These specimens were very old, yet there was no trace of evolutionary change. The Egyptians "have not only left us representations of animals, but even their identical bodies embalmed and preserved in the catacombs" (Cuvier 1813, 123).

If not evolution, then what is the explanation for organic diversity? Where did species come from? Cuvier admitted he did not know. For all that he was a Protestant Christian, he would never have appealed to the Bible as the ultimate source of scientific authority, even though he did indeed think that earthly upheavals (catastrophes, as his English followers called them) occasionally occurred and was prepared to say (on the authority of the Bible taken as a historical manuscript) that Noah's Flood may have been the most recent of these. But beyond allowing that the history of life (as revealed in the fossil record, which he himself was doing much to open up) appears to be roughly progressive, Cuvier had little to say. "I do not pretend that a new creation was required for calling our present races of animals into existence. I only urge that they did not occupy the same places, and that they must have come from some other part of the globe" (Cuvier 1813, 125–126).

Ultimately, however, it was not empirical evidence but final causes that doomed evolution for Cuvier. Following in Kant's footsteps, Cuvier just could not see how change could cross the species barrier without disrupting the organism's ability to survive. The intricacies of adaptation exist specifically to aid the organism in the niche that it occupies. Hybrids are neither fish nor fowl—adapted neither for water nor for air—and as such cannot possibly survive and reproduce.

Homology

Influential though he may have been, Cuvier never had things entirely his own way. Even as he thought and wrote, challengers to his power—if not socially, then intellectually—waited in the wings. By the 1820s Lamarck, now old and blind, had been dismissed with scorn by Cuvier and his supporters, but younger scientists who resented Cuvier's status and authority were not prepared to give an inch.

The key point of dispute was those homologies between animals (and plants) of different species—features apparently built on the same lines or according to the same blueprint (called a *Bauplan* in Germany, an "archetype" in Britain). To many, the similarity was beyond dispute, but from a functional standpoint that similarity seemed worthless as a way to explain origins and purpose. The arm and hand of the human are used for grasping, the front leg of the horse for running, the wing of the bat for flying, the paw of the mole for digging, the flipper of the seal for swimming, and so forth. These ends are often achieved only by riding roughshod over component parts that in other organisms have very definite and distinct functions. The horse has only one toe, whereas humans have five digits, including the crucial reversible thumb. Why the similarity—why the homology—when clearly it has no specific utilitarian end whatsoever?

Cuvier downplayed the similarities, denying absolutely that they exist between embranchments and belittling their significance within embranchments, arguing that they simply represent different although coincidental functional approaches to the problems of life. Of course, the fact is that Cuvier himself was using the facts of homology in his own anatomical reconstructions, whether consciously or not. For all that he was claiming to infer the unknown parts of organisms from the known parts, as often as not—more often than not—he was using his own great knowledge of similar organisms to speculate on the unknown parts of those specimens under study.

Be that as it may, his denials of homology were at odds with the opinions of many of his contemporaries. In Germany a whole new system of morphology—called *Naturphilosophie,* or natural philosophy—was springing up which made homology central. These morphologists saw underlying forms or ideas as informing the structure of organisms, and indeed, with their own links to Kant, the natural philosophers drew strong analogies between the development of the individual and the development of the group or race (which for some meant evolution and for some did not). They believed that the whole of nature is interconnected through shared *Baupläne* and even saw repetition and isomorphism (serial homology) between the parts of the individual organism.

The poet Johann Wolfgang von Goethe was prominent in this school. He reduced the plant world to one basic form—the *Urplant*—and in the vertebrate world he saw everything as a repetition of one basic vertebral piece (Goethe 1946). This supported the so-called "vertebral theory of the skull" which claimed that the skull consists of four or five basic parts, all modifications of the same original vertebral piece.

Closer to home, Cuvier had to contend with a similar, final-cause-belittling stance taken by his fellow French morphologist and sometime friend, Etienne Geoffroy Saint-Hilaire. Ever keen to escape the shadow of his great rival—the romantic Geoffroy had at one point gallivanted off to Egypt with Napoleon while the sober Cuvier stayed home and consolidated his position—Geoffroy found unexpected homologies in the mammalian ear. "Strictly, it will suffice for you to consider man, a ruminant, a bird, and a bony fish. Dare to compare them directly and you will reach in one stroke all that anatomy can furnish you of the most general and philosophical nature" (Geoffroy 1818, xxxviii). Then, generalizing out to other bones in the vertebrate body: "An organ is sooner altered, atrophied, or annihilated than transposed." Initially, Geoffroy did not presume to suggest that connections exist across embranchments, but one thing led naturally to another, and before long isomorphisms were seen across the vertebrate/invertebrate barrier. It drew a predictable response from Cuvier: "Your memoire on the skeleton of the insects lacks logic from beginning to end" (Geoffroy 1820, 34).

The inevitable public confrontation between these two ego-driven biologists took place in 1830 at the French Academy of Sciences. Cuvier and Geoffroy squared off before an excited audience. The ostensible cause of debate was a memoir by two minor biologists who had claimed that, so long as you sufficiently twist and distort the usual living forms of the animals, analogies could be found between the cuttlefish (an invertebrate) and some organisms with backbones. This violation of the embranchments would have been enough to set things off, but there was also a major subtext to the confrontation. Just because one recognizes the significance of homology, one does not automatically become an evolutionist. Aristotle did not, Kant did not, and, although some have claimed otherwise, probably Goethe did not either. But homology is a good piece of evidence—the similarities are due to shared ancestry, and the differences in function are due to changes through time—and it is clear that Geoffroy, who was a friend and admirer of Lamarck (not to mention an ardent progressionist), was inclined that way. Even if he were not convinced on intellectual grounds, the fact that Cuvier was against evolution would have been spur enough. So again, final cause and evolution were put into play against each other.

Geoffroy, as expected, admired the memoir on the cuttlefish. Cuvier, as expected, disliked the memoir—so much so, in fact, that the memoir's authors squeaked that they had had no intention of saying anything which might upset the all-powerful permanent secretary. But everyone had a wonderful time, as the two champions thrust and counter-thrust; the aged Goethe excitedly exclaimed that "the volcano has come to an eruption, everything is in flames."

Shortly after the event, Cuvier died rather suddenly and Geoffroy spent the rest of his life defending at length—at very great length, and with inversely proportional modesty—the significance of his own life's achievements. Owing to their efforts, by the end of the third decade of the nineteenth century, the scientific world knew full well that there is more to organic life than direct adaptation—sometimes called utilitarian adaptation, that is, the end-directed, design-like nature of contrivance (Dawkins's adaptive complexity). Isomorphisms, or homologies as they came to be known, were clearly widespread and in some sense had to be significant.

A PLURALITY OF PROBLEMS

HUMAN BEINGS seem always to throw up exceptions and counterexamples to any generalization, and British thinking on natural theology at the beginning of the nineteenth century fits this pattern. Not everyone was as enthused by the argument from design as was William Paley and the natural theologians who followed in his path. The eighteenth century in Britain (and in America also) had seen great bursts of religious enthusiasm among evangelicals, who believed and preached that reason is not the key to salvation but faith, pure and simple—an opening of one's heart to Christ and to the Holy Word. To people of this faith, arguments for God's existence based on an appeal to design are irrelevant or worse, for they point not to the suffering Lamb of God, who died for the sins of the most lowly, but to an impersonal pagan deity whose nature can never be known by the humble and ignorant.

Low Church Anglicans, similarly minded Presbyterians, and sects, like the Methodists, who broke from the Church of England early in the eighteenth century, spurned the Broad Church tradition of theologians like Paley. Typical was the Scottish preacher Thomas Chalmers, who in 1812 assured his congregation that "all true disciples approached God through the mediation of Christ." He continued: "It is not the God of natural religion that they recognize: that tremendous metaphysical being whom to know is but to tremble. Not the God of the Schools. Not that God which the Fancy of philosophers has formed, and propped upon the feeble imagination of a priori and a posteriori arguments. It is God such as he has chosen to reveal himself. It is the God of the New Testament" (unpublished sermon of 1812, quoted by Topham 1999, 157).

Of course, things are never simple, even for the pure in heart. Enthusiasm for revelation could easily be matched by enthusiasm for natural philosophy. John Wesley, a "young gentleman of Oxford," as he described himself, who had somewhat inadvertently founded Methodism, always had an interest in and sympathy for science, along with an openness to the invention of labor-saving devices and a concern for sensible medical and hygienic practices. Even Chalmers himself later modified his thinking to such an extent that he could write one of the Bridgewater treatises. But there was a wariness about natural theology in Britain, verging on distrust, and by the early nineteenth century Chalmers was far from alone in this feeling.

By one of those interesting quirks of history, this evangelical hostility toward natural theology translated into a major plank of a very different style of religion, one that in the 1830s broke out as surely the most significant force in nineteenth-century British church history—the so-called Oxford Movement. It was centered on a group of churchmen who were determined to reverse the Reformation and to restore the true Catholic roots of the established church. To this end, these High Church Anglicans penned a series of pamphlets *(Tracts for the Times)* in which they defended a vision of Christianity whose continuity was unbroken all the way back to the Church Fathers.

The group's leader—John Henry Newman—decided eventually to convert to Roman Catholicism, and many followed in his wake. As a Catholic convert, Newman had to accept the Thomistic arguments for God's existence, but he never much cared for them (Newman 1870; 1873). Like many in the Oxford Movement, he had grown up in a fervently evangelical home, and wariness about natural theology marked him (and others) for life. In 1870, twenty-five years after he converted, in correspondence about his seminal philosophical work, *A Grammar of Assent,* Newman wrote: "I have not insisted on the argument from *design,* because I am writing for the 19th century, by which, as represented by its philosophers, design is not admitted as proved. And to tell the truth, though I should not wish to preach on the subject, for 40 years I have been unable to see the logical force of the argument myself. I believe in design because I believe in God; not in a God because I see design" (Newman 1973, 97). Making the evangelical's point, he continued: "Design teaches me power, skill and goodness—not sanctity, not mercy, not a future judgment, which three are of the essence of religion."

William Whewell Takes a Turn

We will return to Newman and his fellows, but for now—recognizing that there were varieties of opinion—let us refocus on those directly concerned with the advance of science. With the Napoleonic wars now firmly over, establishment scientists in Britain felt a keen appreciation for Cuvier's contributions to science and sympathy for his attempts to quell the opposition. "The opinions of Geoffroy St. Hilaire and his dark school seem to be gaining some ground in England. I detest them, because I think them untrue. They shut out all argument from *design* and all notion of a Creative Providence" (Clark and Hughes 1890, 2.86). This was one of the milder comments from Adam Sedgwick, a fervent priest of the Anglican Church, professor of geology at the University of Cambridge, and advocate of natural theology. His indignation was not without reason. Although he himself was not working in an area where the facts of adaptation were pressing, some of his clerical brethren were, and every day their findings demonstrated that the tradition of natural theology could pay high scientific dividends to those with energy and imagination.

A classic of the period was the *Introduction to Entomology or Elements of the Natural History of Insects* (1815–1828), co-authored by the Reverend William Kirby, vicar of Barham. Dealing with a topic of much general interest—amateur collectors read the volumes with keen pleasure and profit, as did those (like beekeepers) who were professionally interested in the subject—Kirby displayed and discussed the facts of organized complexity and linked them directly and un-self-consciously to the Creator of all things. In his discussion of means of defense, we learn that for insects it is "their colour and form, by which they either deceive, dazzle, alarm or annoy their enemies" (Kirby and Spence, 1815–1828, 2.219). Camouflage is a major tactic. "Thus one of our scarcest British weevils (*Curculio nebulosus,* L.), by its gray colour spotted with black, so closely imitates the soil consisting of white sand mixed with black earth, on which I have always found it; that its chance of escape, even though it be hunted for by the lyncean eye of an entomologist, is not small." But do not think that the causes behind any of this are at all in doubt. "One thing is clear to demonstration, that by these creatures and their instincts, the power, wisdom and goodness of the GREAT FATHER of the universe are loudly proclaimed" (2.217). For good measure, we have "the atheist and infidel confuted" and "the believer confirmed in his faith and trust in Providence." One cannot ask for much more than that.

Except perhaps, looking back from the perspective of two centuries, one might like some evidence that even the most confident of British natural theologians were starting to realize that simple appeals to adaptation and design were no longer enough. Even if good Britons were not about to be seduced by vile German or French idealisms, one might like some indication that they saw more in nature than simple end-directed complexity and that the various problematic phenomena so evident in the living world were not to be ignored. Of course, some nonfunctional patterns could be simply ascribed to the Creator's good taste, which apparently ran along the lines of the taste of the average housewife. For as the religious polemicist and amateur geologist Hugh Miller (1856) was happy to note, the Lancashire calico pattern called "Lane's Net" was used by the Almighty many years before when he made some of the corals now to be found fossilized in the Old Red Sandstone of Scotland. "The beautifully arranged lines which so smit the dames of England, that each had to provide herself with a gown of the fabric which they adorned, had been stamped amid the rocks *eons* of ages before" (p. 242).

But this kind of whimsy was hardly adequate for those seriously trying to make sense of the analogies and isomorphisms highlighted by the *Naturphilosophen.* Kirby realized that these homologies could not be ignored, and he was one of many who became enamored with a home-grown idealistic system, the so-called quinary system of William MacLeay (1819–1821). MacLeay saw nature as arranged in rings of similar species that then could be grouped by fives into ever-larger touching ("osculating") circles, and so on up. Strange as such a system may seem to us now, it helped to make sense of the various isomorphisms to be found between different species, and it fit with the growing British enthusiasm in the nineteenth century for Greek thought, particularly that of Plato.

This quinary system was regarded as no less a mark of God's design intentions than were adaptations. It showed how he was planning the world according to fixed rules—based on the perfect figure, the circle, no less. Kirby felt so little conflict with his general appeal to adaptation that in a later edition of the *Introduction to Entomology* (1825), in a short history of the subject, he included a section "The Era of MacLeay." This certainly did him no harm in the eyes of the general public, for a year or two later Kirby was invited, like the Reverends William Buckland and Thomas Chalmers, to write one of the Bridgewater treatises.

They were joined in this task by Sedgwick's great friend, the sometime professor of mineralogy at the University of Cambridge and fellow member of Trinity College, William Whewell (pronounced "Hule"). Also an enthusiast for natural theology—for Cuvier's work in particular—Whewell was by far the most sophisticated thinker on these subjects in the years leading up to Charles Darwin's *Origin of Species.* At first his contribution to the topic had to be a little circumspect. In his Bridgewater treatise, Whewell could not plunge right into a hymn to the magnificence of organized organic complexity, for the assignment he had drawn was *Astronomy and General Physics* (1833). Prima facie, the organic world was not his direct field of interest and discussion. He was stuck with what was known as the cosmic argument (design from the whole of nature), rather than what was known as the physico-theological argument (design inferred from organisms specifically). However, showing the talents that were to move him rapidly to a place of intellectual prominence among his countrymen, Whewell—as much a philosopher as a scientist (later he was to become Knightsbridge Professor of Philosophy at Cambridge)—brought off his task with stunning success. His contribution to the series was truly of far higher quality than those of most of the other writers, particularly the rather crude speculations from the pen of Buckland.

Ever wary of argument-stopping counter-moves, Whewell began by making himself safe from those who would throttle him with Humean critiques. He was careful to start his Bridgewater treatise with a disclaimer about the relative significance of natural theology. It must always, we learn, be secondary to revelation. His labors were to show how the world "harmonizes with the belief of a most wise and good God" (Whewell 1833, vi). One can do this much, but no more. "This, and all that the speculator concerning Natural Theology can do, is utterly insufficient for the great ends of Religion; namely, for the purpose of reforming men's lives, of purifying and elevating their characters, of preparing them for a more exalted state of being. It is the need of something fitted to do this, which gives to Religion its vast and incomparable; and this can, I well know, be achieved only by that Revealed Religion of which we are ministers" (pp. vi–vii).

This was not simply philosophical chicanery. Whewell was a practicing Christian and, owing to his position as fellow of a Cambridge college, an Anglican priest. But he always took Jesus' exhortation to meekness metaphorically rather than literally, and he certainly was not about to let a Scotch skeptic, whatever his qualifications, silence him before he began. Absolute proof may be impossible, but there is still much to be said on the natural theological front.

For a start, Whewell thought that all laws of nature, in themselves, attest to God's existence and greatness. Moreover, he thought that the very existence of universal laws fits in with God's purpose for us here on earth. We humans are not put here simply to exist. We have tasks, trials, imposed by God. These trials are not just moral but intellectual, for it is our task to trace the motions of the heavens, whether or not they have any direct utilitarian purpose. "The contemplation of the material universe exhibits God to us as the author of the laws of material nature; bringing before us a wonderful spectacle, in the simplicity, the comprehensiveness, the mutual adaptation of these laws, and in the vast variety of harmonious and beneficial effects produced by their mutual bearing and combined operation" (Whewell 1833, 251).

Yet, this all said, even Whewell had to cheat a bit. Whatever his given topic, ultimately his job under the Earl's bequest was to promote and garnish the argument from design. Given the weakness of the evidence in the nonliving world for the argument to organized complexity—physical laws just do not cut the mustard like adaptations—he had somehow to bring things around to the living world, the world of animals and plants. That is, he had to slide from the cosmic argument to the physico-theological argument. Fortunately, as Whewell demonstrated at length—with many detailed examples—this turn can be achieved because the nonliving world is designed as a home for the living world, which conversely accommodates itself to the nonliving world in turn. What if the length of the year were more or less than it is? Growing seasons would collapse in chaos. What if plants and animals did not have the exact breeding cycles that they do? Again there would be chaos. Such coincidences were no coincidence. They had to be designed.

Whewell is sensitive to the Humean argument that the world is not truly analogous to a product of design and hence we cannot legitimately argue to an end-directed organized complexity, and from there to a designer behind it. Like Paley and everyone else at the time, Whewell thought that ultimately one has to argue for a designer because there is no other option—in the presence of organized complexity, a designer is simply the best explanation. But he did think that the issue had to be tackled head-on, and so, trying to short-circuit the Humean objection, Whewell (in a very Kantian-sounding argument) argued that there is in fact no inference here. We do not argue the case for organized complexity and then the case for final cause; rather, we assume final cause as a condition of experience at all.

In Whewell's account, the critics take this position: "The universe, considered as the work of God, cannot be compared with any corresponding work, or judged by any analogy with known examples. How then can we, in this case, they ask, infer design and purpose in the artist of the universe? On what principles, on what axioms can we proceed, which shall include this necessarily singular instance, and thus give legitimacy and validity to our reasonings?" (pp. 343–344). In response Whewell argues: "When we collect design and purpose from the arrangements of the universe, we do not arrive at our conclusion by a train of deductive reasoning, but by the conviction which such combinations as we perceive immediately and directly impress upon the mind." For those who recognize this, no argument is necessary, and for those who fail to recognize this, no argument is possible.

Fundamental Ideas

As it turned out, the Bridgewater treatise was but a preliminary. Later in the decade, Whewell produced two massive works for which he is justly famous: in 1837 *The History of the Inductive Sciences* in three volumes and then in 1840 *The Philosophy of the Inductive Sciences* in two volumes. Given that his thinking stressed the significance of law in itself, one move Whewell might have made would have been to reinforce this, in the way that Baden Powell was to do a few years later. He could have made much of the ability of God to build design right into the workings of law itself. In an unofficial addition to the Bridgewater treatises, Charles Babbage (1838), designer of a proto-computer, had shown how he could set his machine to produce any exceptions he pleased, to any regular series whatsoever—a series of natural numbers up to 1,000,001 for instance, and then 1,010,002 as the next number in the series, or any other number for that matter. Law could be jiggled around to produce design, and this in itself would be proof of God's existence—no need to kill yourself finding and stressing complexity.

But perhaps sensing that something like this might be the thin edge of the wedge for a completely naturalistic approach to the world, Whewell took a very different tack. Now explicitly showing the Kantian influence, although with a growing sympathy for a kind of Christian Platonism (where God himself stands directly behind both the world itself and the way in which we interpret the world), Whewell argued that certain fundamental ideas structure our science. We impose this understanding on the facts of experience, and this transforms our knowledge from random sensation to an informed, connected framework of necessary truths. "I use the term *Idea* here to designate those inevitable relations which are imposed upon our perceptions by acts of the mind, and which are different from anything which our senses directly offer to us" (Whewell 1840, 1.26).

What makes Whewell's philosophy interesting and distinctive is the fact that he saw different fundamental ideas as being operative in different sciences. We start in the physical sciences with space and time and cause and so forth, and then move over to more specialized ideas in more specialized sciences. In the organic world, Whewell thought the distinctive notion was that of organization, and he quoted Kant's claim that organization involves all of the parts being "mutually ends and means" (2.77). However, there has to be more than just this (2.78):

> Stones slide from a rock down the side of a hill and cause it to be smooth; the smoothness of the slope causes stones still to slide. Yet no one would call such a slide an organized system. The system is organized, when the effects which take place among the parts are *essential to our conception of the whole*; when the whole would not be a whole, nor the parts, parts, except those effects were produced; when the effects not only happen in fact, but are included in the idea of the object; when they are not only seen, but foreseen; not only expected, but intended: in short when, instead of being causes and effects, they are *ends* and *means*, as they are termed in the above definition.
>
> Thus we necessarily include, in our Idea of Organization, the notion of an end, a purpose, a design; or, to use another phrase which has been peculiarly appropriated in this case, a *Final Cause*. This idea of a Final Cause is an essential condition in order to the pursuing our researches respecting organized bodies.

In other words, Whewell argued that final cause is something we impose on organic nature in order to understand it. We cannot separate the organized complexity of the living world from our interpretation of it as something end-directed. This is a Kantian position that goes further, in that Whewell was quite open in thinking that God stands behind final cause—also, unlike Kant, Whewell did not just think that final cause is a necessary heuristic for understanding nature. He wanted to elevate it to the same status and the same necessity as Newton's causal laws of motion.

Against Evolution

Whewell's opposition to evolution was overdetermined—that is, it had more causes than were required to produce it. For a start, as an ordained minister in the Church of England, he saw evolution as incompatible with his Christian faith. He would have scorned the need to read Genesis literally, but he certainly saw the notion of progress in evolution as incompatible with the Providentialism he detected in human history. More than this—another counter against progressionism—politically Whewell moved from being a Whig (a liberal) to being a conservative, and early in the 1840s became master of his college, the most powerful in Cambridge. Living on the rents of the houses and farms owned by Trinity, Whewell had little enthusiasm for ideologies of change, whatever the cause of it.

Then there was the vital question of the standing of science in early Victorian Britain. Whewell was at the front of those committed to making a place for natural philosophy within and without the university: he was much involved in the scientific societies of the day (like the Geological Society of London and the newly founded British Association for the Advancement of Science); he was himself, through his writings (the *History* and the *Philosophy* particularly), trying to articulate the norms of quality science; and he was encouraging and mentoring young scientists, notably (in hindsight) Charles Darwin. This was reason enough to recoil from notions of evolution, which had the taint of the second-rate. With every good reason, transmutationism (as evolution was called in Britain for a time) was on a par with quasi- or pseudo-sciences like phrenology, the study of cranial bumps. (Robert Chambers, significantly, had set out to write a book on phrenology but changed his topic to evolution halfway through.)

Third were all of the empirical arguments of Cuvier, whom Whewell justifiably took to be the authority on these issues. And included in these—if Whewell needed any encouragement—was the problem of final cause. Most revealing on this score was the fact that in 1845, in response to *Vestiges,* Whewell collated extracts from earlier writings in one little volume, *Indications of the Creator.* The book was stuffed with material about final cause: "Recognition of final causes in physiology," "Use of final causes in physiology" (from the *History*), and "The idea of final causes" (from the *Philosophy*). And more. All in all, as the great Frenchman had shown: "Indefinite divergence from the original type is not possible; and the extreme limit of possible variation may usually be reached in a short period of time: in short, *species have a real existence in nature,* and a transmutation from one to another does not exist" (Whewell 1837, 3.576). What then are we to say about origins? "The mystery of creation is not within the range of her legitimate territory; she says nothing, but she points upwards" (3.588).

This was hardly satisfactory in mid-nineteenth-century England; giving everything over to God was no longer an option for a scientist. And what about the problems with final cause? What, in particular, about the threat from homology? Frankly, if he had had his druthers, Whewell would probably have ignored it entirely, in all its foreign and home-grown varieties (*especially* the home-grown varieties, after Chambers made the quinary system a lynchpin of his evolutionary system). But silent contempt for such phenomena as homology just would not suffice.

The leading anatomist of the day, Richard Owen, Hunterian Professor at the Royal College of Surgeons, was a good friend of Whewell (they were former school classmates) and his unofficial scientific adviser on the life sciences. Owen, who was dubbed the "British Cuvier," stressed adaptation. Whewell was ever fond of quoting an article by Owen on the way in which kangaroo mothers feed their near-embryonic young, ensuring that milk will be ingested and choking avoided. It is, Owen (1834) said and Whewell echoed, "irrefragable evidence of Creative forethought." However, adaptation is not enough. Owen, who had been to France and mixed with people like Geoffroy and who had a yen for the speculations of the *Naturphilosophen,* also stressed homology. Indeed, he coined the term, showing how crucial a notion he thought it to be. Far too crucial to be ignored by his chum Whewell.

And so, at first, homology was recognized, albeit in a somewhat ungracious way, prior to being downgraded. "Whenever such laws are discovered, we can only consider them as the means of producing that adaptation which we so much admire. Our conviction that the Artist works intelligently is not destroyed, though it may be modified and transferred when we obtain a sight of his tools. Our discovery of laws cannot contradict our persuasion of ends; our Morphology cannot prejudice our Teleology" (Whewell 1840, 2.88–89). But this was really not quite good enough either. Whewell may not have been a practicing anatomist, but Owen was, and he knew that homology is too significant a factor to be brushed aside in this way. Hence, through the 1840s Owen developed and articulated a position that takes account of both teleology and morphology (speaking, in the sense of Whewell, as if these two were opposed).

Concentrating on the vertebrates, which was his particular area of expertise, Owen argued that we can puzzle out and specify the actual blueprint or archetype on which all vertebrates are modeled. It is a fish-like form, which (in the tradition of the *Naturphilosophen*) shows serial homology from top to bottom between its parts. Adaptation is overlaid on this, thus making particular kinds of vertebrates.

By this time in Britain, Platonism had become a major intellectual movement, and in Cambridge, where Whewell was now both professor of philosophy and master of Trinity, he was the dominant educational force. As such, Whewell argued strongly for the Platonic ideals in education—especially since they put heavy emphasis on the significance of mathematical training. Picking up on this trend, at times Owen seemed to endorse the Whewellian claim about the idea of final cause: "The platonic idea or specific organizing principle would seem to be in antagonism with the general polarizing force, and to subdue and mould it in subservience to the exigencies of the resulting specific form" (Owen 1848, 171). At other times, however, he seemed to suggest that the archetype is the Platonic Form. "What Plato would have called the 'Divine idea' on which the osseus frame of all vertebrate animals . . . has been constructed" (Owen 1894, 1.388). No matter really. The point is that by mid-century it was realized by the leading authorities that scientifically one had to go beyond—or at least supplement—the traditional Aristotelean/Cuvierian stance on final cause.

Plurality of Worlds

Whewell recognized this. The third edition of the *History* (1857) has an appendix where Owen's line is accepted completely, subject to the reservation that final cause is being constrained, not destroyed. "The arm and hand of man are made for taking and holding, the wing of the sparrow is made for flying; and each is adapted to its end with subtle and manifest contrivance. There is plainly Design" (3.559). Yet undoubtedly the argument in favor of organized complexity had been knocked sideways. By the 1850s, it was, paradoxically, Whewell himself who was putting pressure on the argument from complexity to design. Writing anonymously (because it would have been unseemly for the master of Trinity to be involved in open controversy), but with an identity known to all, Whewell's publication *Of the Plurality of Worlds* showed that natural theology was under pressure, and not because he had lost his faith—the very opposite in fact. Whewell was rather showing that, in the light of modern science, juggling both revealed theology and natural theology is no easy thing—there are tensions and contradictions.

The spur for this crisis came about through renewed debate on an age-old topic, the possible existence of intelligent beings on heavenly bodies other than the earth. Is there indeed life teeming throughout the universe? Or is the vast, deep space entirely barren, with stars circling endlessly in a dead vacuum? The popular assumption—one to which Whewell himself had earlier subscribed in his Bridgewater treatise—was that there must be life, including intelligent life, elsewhere in the universe. Otherwise everything would be so pointless. To have made the whole universe with but us humans would be a violation of God's own good sense. It would conflict with the notion of a careful, designing creator. It would be without purpose.

This had also been the position of Thomas Chalmers, who recognized nevertheless that there would be theological challenges, which he was prepared to answer in a distinctively Presbyterian way. Asked the psalmist: "When I consider the heavens, the work of thy fingers, the moon and the stars, which thou has ordained: What is man, that thou are mindful of him? and the son of man, that thou visitest him?" Chalmers's answer was that modern science clearly points to a creation teeming with life. Our planet is too similar to other possible abodes seriously to think we may be alone. Living beings—intelligent living beings—are everywhere. Moreover, not only are they (like us) made in the image of God, but they know and cherish this fact. "Is it presumptuous to say, that the moral world extends to these distant and unknown regions? that they are occupied with people? that the charities of neighborhood and home flourish there? that the praises of God are there lifted up, and his goodness rejoiced in?" (Chalmers 1817, 44). What then of us humans down here on earth? Can God possibly care about us? He most certainly does! The "divine condescension" reaches out to us, insignificant and worthless as we are. Indeed, the paradox is that precisely because we are so undeserving we can catch a glimpse of God's power, love, and grace. The very multiplicity of life confirms the greatness of God and the significance of our bond with him.

This kind of extreme Protestantism, which puts everything on the whim of God—"justification by grace"—never sat that well with Anglicans, with their regard for the value of good works (known technically as the heresy of Pelagianism). Certainly self-abrogation was not to the taste of one like William Whewell, who regarded himself and his achievements with utter complacency. (Wags used to joke that Whewell wrote on the plurality issue to show that "through all infinity, there is none so great as the Master of Trinity.") Even in the Bridgewater treatise of 1833, Whewell regarded God somewhat as a tutor of Trinity, who sets the discerning of his laws as the fellowship examination of life. The student is supposed to succeed through talent and hard work, not through the favor of the professor. By the 1850s, Whewell was breaking from Chalmers on the science and natural theology also.

Why? On the one hand, Whewell felt the need to preserve the unique status of humankind. It is human beings here on earth who have a special relationship with God—it is we who are made in his image—and it is for our sins alone that he died on the cross that we might be saved. The crucifixion, the conquering of sin, the resurrection have to be for our ends only. We cannot have Jesus dying on the cross, sequentially through time and space. "The earth, thus selected as the theatre of such a scheme of Teaching and Redemption, can not, in the eyes of anyone who accepts this Christian faith, be regarded as being on a level with any other domiciles. It is the Stage of the Great Drama of God's Mercy and Man's Salvation; the Sanctuary of the Universe; the Holy Land of Creation; the Royal Abode, for a time at least, of the Eternal King" (Whewell 1853, 44). One can hear the echoes of the Master's Sunday chapel sermons to undergraduates.

On the other hand, as Whewell saw it, a full-blooded plurality begs for—practically necessitates—a background of evolutionism. Instead of organisms appearing providentially here on Earth—coming through miraculous interventions of a benevolent Creator—they arrive apparently to order, whenever and wherever space is available. If this is not to come by law, in an evolutionary fashion, it is certainly to come by some-

thing suspiciously similar. In Whewell's view, such a horrendous possibility must be stopped in its tracks before it can take its flight of fancy. Hence, he argued strenuously that science, properly understood, suggests strongly that no life exists elsewhere in the universe, certainly no human-like life.

In an argument that equally scandalized and titillated the imaginations of his fellow Victorians, Whewell paid special attention to the barren deserts of the rest of our solar system. He was going to have no truck with little green men from Mars or from anywhere else. Jupiter is little more than solid ice and Saturn not much better. Except of course, being so big, gravity would be colossal. "For such reasons, then, as were urged in the case of Jupiter, we must either suppose that it has no inhabitants; or that they are aqueous, gelatinous creatures; too sluggish, almost, to be deemed alive, floating in their ice-cold waters, shrouded for ever by their humid skies" (Whewell 1853, 185–186).

But is the rest of the universe therefore without a point, without any sense of existing for or being directed toward a valued end? Can it be, as Whewell's friend, Sir James Stephen (the Regius Professor of History at Cambridge) put it, that "the measureless shoreless ocean of space enveloping us, embraces no one world sinless, wise, holy, happy? no region in which the all seeing eye can rest with complacency, or of which the Divine voice can once again declare that 'it is very good'?" (letter to Whewell 15 October 1853, Whewell Papers, Add Ms a 216.130). Cleverly making a virtue of necessity, Whewell turned organic homology to good account. Just as morphology shows that not everything has a direct function, so it is not necessary to assume that the whole of cosmological creation has a direct function. "Instead of manufacturing a multitude of worlds on patterns more or less similar, He has been employed in one great work, which we cannot call imperfect, since it includes and suggests all that we can conceive of perfection. It may be that all of the other bodies, which we can discover in the universe, show the greatness of this work, and are rolled into forms of symmetry and order, into masses of light and splendour, by the vast whirl which the original creative energy imparted to the luminous element" (Whewell 1853, 243).

In arguing thus, Whewell was treading on thin ice, and he knew this. Once you start promoting the significance of homology, can evolution be far behind? This was precisely the move made by Chambers in *Vestiges* and earlier by Geoffroy, and this was certainly hinted at (and more, in some cases) by the *Naturphilosophen*, especially given their fondness for analogies between individual development and group development. Nonfunctional homology especially points to common origins, and how more easily can one get from these origins than by evolutionary change? Owen was already coming under severe scrutiny from people like Sedgwick, who feared that homologies and archetypes formed a slippery slope down to transmutation. (They did! Owen had already sent a private fan-letter to Chambers.) So this line of argument had to be stopped cold. Specifically, "the evidences of design in the anatomy of man are not less striking than they were, when no such gradation was thought of. And what is more to the purpose of our argument, the evidences of the peculiar nature and destination of man, as shown in other characters than his anatomy,—his moral and intellectual nature, his history and capacities,—stand where they stood before; nor is the vast chasm which separates man, as a being with such characters as these latter, at all filled up or bridged over" (Whewell 1853, 216–217).

"His moral and intellectual nature, his history and capacities." This had been the starting point for the argument of the Bridgewater treatise—the task set by Whewell's Anglican God, for men of intelligence and diligence. Twenty years later, he again makes his appearance, as Whewell turns now to argue that law itself is a mark of God's activity and magnificence, and that the discernment of this law is our task here on earth.

Whewell's *Essay* sparked a huge controversy. My favorite response—the magnificently titled *More Worlds than One: The Creed of the Philosopher and the Hope of the Christian*—came from Sir David Brewster, Scottish man of science and ardent Presbyterian. "The chariots of flame and the horses of fire that bore Elijah from his star of earth, and surrounded Elisha in the mountains of Syria, and the wheels of amber and of fire that were exhibited to the captive prophet on the banks of the Chebar, become, in the poet's eye, the vehicle from planet to planet, and from star to star, in which the heavenly host is to survey the wonders and glories of the Universe" (Brewster 1854, 261). And so on and so forth. Intelligent life was to be found everywhere, including the surface of the sun. But we need not follow the details here, for the points have been made.

One of the many things that Thomas Kuhn said in his *Structure of Scientific Revolutions* was that old theories or paradigms do not simply get pushed aside by the merits of the new. They fail in their own right, collapsing into disarray and paradox. They open up a space that needs to be filled. This surely seems to be the case here, as one looks at the thinking of Whewell, the most subtle of the Victorian philosophers on matters of teleology, organic nature, and natural theology. He is forced to modify—circumscribe—the argument for organic complexity dramatically. It still stands, and is very significant, but it is not all-embracing, and the exceptions cannot simply be dismissed and ignored. They must be addressed. The argument from complexity to a deity is in worse disarray, even if one puts to one side the Humean objections. The exceptions like homologies need explanation, and it is less than plausible simply to say that God is a Platonist and patterned everything on archetypes—at least, this response asks as many questions as it answers. As Whewell surely shows, the entire argument from design sets up tensions with revealed theology that need answering—in Whewell's fashion or in some other way. Do we see value here on earth in the inorganic domain? Do we see value here on earth in the organic domain? And if so, value to whom? And is there value elsewhere in the universe, and if so what is it and for whom? And if no value, why did the extended universe get created at all?

Above all, however one solved the value issue, there was the worrying mix of science and theology. The argument to design gave an answer to the question set by the argument to complexity, but it was not a scientific answer—and that increasingly was becoming a problem. It was especially a problem for someone like Whewell who was, on the one hand, pushing science and the rule of law for social and professional reasons and who was, on the other hand, pushing the rule of law for natural theological reasons. One can hardly then sit back and say, without raising eyebrows, that God set as our task here on earth the finding of the laws by which he created, and then turn around and say that when it comes to some of the most interesting creations of all, he did not use law! And yet, how do you get out of this dilemma? If you start talking about law doing things on its own, you run into all of the problems raised by people like Kant and Cuvier. Blind law just does not lead to contrivance. Unless you give it all away to people like Babbage and Powell, and then you are virtually on the step of—or have stepped over into—evolution.

All in all, a fresh approach was needed, and this we shall encounter in the person of a young barnacle expert by the name of Charles Darwin.

CHAPTER 5

CHARLES DARWIN

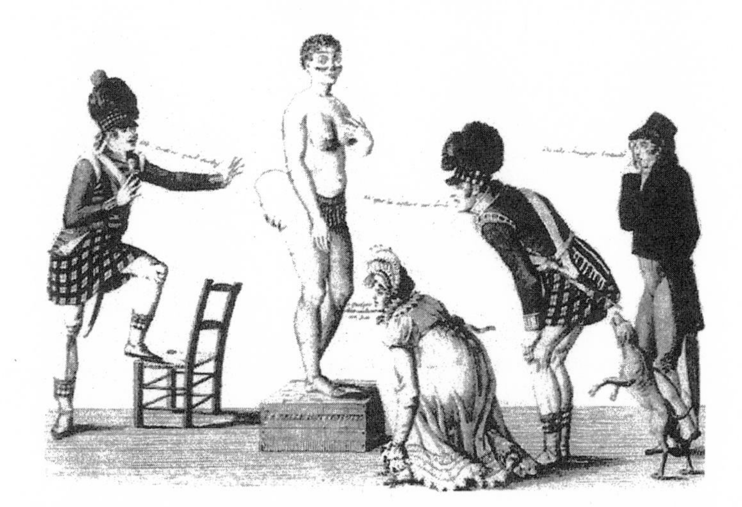

OPINION IS DIVIDED. Darwin's great English supporter, Thomas Henry Huxley (1864), wrote: "That which struck the present writer most forcibly in his first perusal of the 'Origin of Species' was the conviction that Teleology, as commonly understood, had received its deathblow at Mr Darwin's hands" (p. 82). Darwin's great American supporter, Asa Gray (1876), however, wrote of "Darwin's great service to Natural Science in bringing it back to Teleology; so that, instead of Morphology *versus* Teleology, we shall have Morphology wedded to Teleology" (p. 237). Today, we find the morphologist Michael Ghiselin (1984) writing that, far from preserving teleology, "in any nontrivial sense of that word, he did the exact opposite, getting rid of teleology and replacing it with a new way of thinking about adaptation" (p. xiii). To the contrary, the historian James Lennox (1993) responds that in Darwin's work one finds an explanatory structure that is "irreducibly teleological" (p. 418). And Darwin himself is no great help. He thought that Gray's thinking on the subject verged on taking evolutionary biology out of the range of genuine science (Darwin and Seward 1903, 1.191). Yet he praised Gray, telling him that "what you say about teleology pleases me especially" (Darwin 1887, 3.189).

Let us start at the beginning with Charles Darwin himself and then, in the next chapter, see how teleology impinges—if it does—on his science. And how his science impinges—if it does—on teleology.

Charles Robert Darwin (1809–1882) was the second son of Dr. Robert Darwin, oldest son of Erasmus Darwin and a very successful physician in the British Midlands (like his father). His mother was the daughter of Erasmus Darwin's industrialist friend Josiah Wedgwood, who was responsible for the development of the pottery trade in late-eighteenth-century Britain. Not only did Charles inherit good, solid, upper-middle-class status from both sides of the family, but he could also draw on a considerable fortune to ease life's way. At the age of thirty, Charles married his first cousin Emma Wedgwood, so even more family cash (reinforced by successful, British, middle-class values) became available.

This point is worth stressing for two reasons. First, because we should realize that Darwin always had the support he needed to follow his own interests—he never held down paid work, with an agenda dictated by others. Moreover, he had the self-confidence of one for whom nature's bounty had been abundant from the earliest years. Second, as one who was doing so well during the Industrial Revolution and enjoying the comfortable status that it ensured, we should not expect to find young Charles Darwin to be a rebel, forever kicking against the pricks and trying to break from his past. Why would he? What we would expect to find—as indeed we do find—is that, inasmuch as Darwin was a revolutionary (and I do believe he was a very great revolutionary), he would become so because he took the parts given to him and accepted by him and rearranged them into an altogether new picture. Darwin the scientists would not imitate the Christian God, making things from nothing. Rather, as a kaleidoscope rearranges the parts to make a new pattern, so Darwin would rearrange his parts to make his world anew.

Appropriate to his station, Darwin was sent first to one of England's great private schools (misleadingly known as public schools), and then at the age of seventeen to Edinburgh University to train in the family tradition of medicine. He was not a great academic success, and after two years at university he dropped out—bored with the lectures and revolted by surgery. In his *Autobiography*, Darwin wrote in a deprecating way of his early talents and achievements, but this is all a little misleading. Because school education in those days was confined virtually exclusively to the classics—no modern languages, no history or geography, and above all no science—anyone with empirical leanings would truly be a square peg in a round hole. We know, however, that Charles and his older brother (another Erasmus) had a keen interest in chemistry—very much the industrial science of the day—and carried through quite elaborate experiments at home. We know also that, at Edinburgh, Charles started to mix with scientists, especially naturalists interested in the living world. One of his acquaintances was Robert Grant, an anatomist and avowed evolutionist. So quite apart from his grandfather's work (the young Charles read Erasmus's major treatise, *Zoonomia*), evolution was an idea to which Charles Darwin was introduced at an early age.

Acquaintance notwithstanding, Charles certainly did not immediately embrace evolution. To the contrary, the youthful Darwin accepted in fairly literal form the whole of Christianity, including the early chapters of Genesis, and this was a factor in his redirected choice of careers: to be an ordained minister in the established Church of England. To achieve this end, one needed a degree from an English university; and so, at the age of nineteen, in 1828, Charles Darwin was packed off to Christ's College, Cambridge. Formally, his new career followed the same somewhat undistinguished path as the old, and informally he exhibited the same fascination with the world of experience, especially the world of living beings.

Eventually Darwin attracted the attention of others at the university who were interested in science—John Henslow, professor of botany, Adam Sedgwick, professor of geology, and William Whewell, professor of mineralogy and, as we have seen, authority on everything. There were no degree-granting programs on any of these subjects, but for three years Darwin attended systematic lectures on botany given by Henslow. Also, at the end of his degree time he went for a *de facto* crash course on geology with Sedgwick, and he was much in the company of Whewell, who would have lectured him (as he did everyone else) on many topics, especially scientific methodology. ("Science was his forte, omniscience his foible," quipped the wit Sydney Smith.)

The Beagle Voyage

It was through these men, Henslow particularly, that Darwin got his big break. Graduating in 1831, he put his career as a clergyman on hold by accepting an offer of a lengthy voyage on *HMS Beagle,* which was just about to begin a surveying trip around South America. Starting off as traveling companion to the captain, Darwin soon evolved into the ship's naturalist, making large collections of plants and animals and rocks and other objects of interest, all of which were packed up and sent back to

England for classification and study. This trip was to spell the end of Darwin's clerical aspirations. The voyage lasted until 1836, and on his return to England Darwin plunged at once into the life of a full-time scientist (supported by family money), first working on his collections and then spreading out to other interests, including—in the 1840s and '50s—a several-year-long study of barnacles.

Not that Darwin set out from England with the intention of avoiding a clerical life. At the beginning of his voyage, Darwin had no thoughts of changing career plans yet again, and his literal approach to Christianity persisted. This would have been reinforced by his years at Cambridge, where one of the significant books he read was Paley's *Evidences of Christianity.* Dealing with issues of revealed religion, Paley argued that the Gospels are reasonable because of the miraculous events described there and the reliability of the disciples who witnessed them—the future careers of these men, including martyrdom for their faith, was evidence enough of their veracity. At this time, therefore, Charles Darwin's religion was not just one of faith but one where supernatural intervention was absolutely central.

This all took a significant knock on the *Beagle* voyage. As he set off, Darwin was given the first of the volumes of Charles Lyell's *Principles of Geology* (1830–1833), and the other two were sent out as they were published. In this work, Lyell opposed the catastrophism of Cuvier, which had been endorsed with enthusiasm in England by Sedgwick and others. Against the picture of a world cooling from an initial heated state, punctuated by massive upheavals or catastrophes, with organic life being swept away and having to be re-created or imported each time anew (and somewhat different from before to suit the new climate), Lyell argued that earth's history is practically without limit, that overall it has enjoyed a kind of steady state, with fluctuations in heat but with no direction, and that the ordinary causes we see around us—rain, snow, cold, earthquakes, volcanoes, and the like—are quite enough to bring about all of the effects we can read from the geological record. These causes, moreover, have worked throughout history at the intensities that we see around us today. The creation of new species, along with corresponding extinction, is a natural, not a miraculous, event that has continued from time immemorial.

Darwin found Lyell's uniformitarianism (as Whewell was to christen it) very seductive, and he became an ardent enthusiast and exponent of this view. He was particularly concerned to find confirmatory evidence for what Lyell labeled his "grand new theory of climate," the mechanism by which supposedly one gets fluctuations in temperature without overall direction. Lyell argued that this is a by-product of the ever-changing distributions of land and sea—the Gulf Stream shows that temperatures are much more than simply relative distances from the equator—and that these distributions are a function of the ever-rising and ever-falling of the earth's surface. For Lyell, the globe is like a giant water bed, and as deposition (say, from rivers) causes one part of the surface to fall, so correspondingly another part must rise.

Darwin bought into this picture completely, and it became the foundation of his theory of coral reefs, a geological contribution which first persuaded people that here was a person with scientific ability above average. Darwin argued that the circular reefs of coral in tropical seas, sometimes standing alone and sometimes encircling an island, are a result of the falling seabed. Coral, a living creature, can flourish only at the water's surface. As the base beneath sinks, therefore, the coral grows upward and the reefs are created. This was a theory inspired by Lyell and in turn confirmatory of Lyell.

Darwin's ardent enthusiasm for uniformitarianism had two important consequences for his religious outlook. First, the Lyellian theory obviously minimizes miracles, which are a mainstay of Christian belief. In accepting uniformitarianism, Darwin rejected miracles. And as the miracles went out the window, so did Christianity, never to return. Yet Darwin did not plunge straight into a denial of God's existence. I doubt he ever became an atheist, and even toward the end of his life came only to a somewhat tempered agnosticism. On the *Beagle* voyage and right through the writing and publication of the *Origin of Species,* Darwin continued to believe in the existence of God. For Darwin, as for his grandfather Erasmus before him and indeed for Lyell, his theological commitment was to deism rather than to theism. He grew to accept an Unmoved Mover who works through unbroken law, rather than a God of intervention who works through miracles that break physical laws. For Erasmus Darwin and for Lyell, the greatness of God lay in his ability to plan everything beforehand and then just step back and watch it all unfurl as he intended. This was the God that Charles Darwin accepted. The functioning of universal law was proof of his existence and magnificence rather than an embarrassment to be explained away.

Although Darwin's abandonment of Christianity was clearly a significant move theologically, in other respects (socially in particular) it was really not that iconoclastic. The Darwin family was Anglican, and this was the religion in which Charles had been raised. But deism lurked in his background, among the Wedgwoods as well as the Darwins. The Wedgwoods were prominent Unitarians, meaning that they believed in a God who takes an interest in the welfare of individuals but they denied the divinity of Christ and the veracity of the Biblical miracles. Particularly on this side of the family, therefore, Charles was fitting into the pattern rather than breaking it.

So the first point where Lyell was significant for Darwin was in helping him move to a religion where evolution would be a confirmation rather than a refutation of the Divine. The second point centered on science rather than theology. To prove his theory of climate, Lyell had to search for indirect evidence. One important source lay in the distributions of animals and plants. Lyell thought that (nonmiraculous) creation and extinction were continuous and that, however creation came about, the successors were only moderately changed from the predecessors. Hence, by looking at geographical distributions of organisms, one should be able to discern past geological events.

Suppose you find two populations separated by a natural barrier like a mountain range. If they are much alike, one can infer that the barrier is relatively recent. If they are much different, you can infer that the barrier is ancient. Darwin, as an ardent Lyellian, was therefore sensitized to the significance of geographical distributions. It was something he was interested in and which he would investigate as far as he could, whenever the *Beagle* touched shore.

Coming to Evolution

In 1835, leaving South America, the *Beagle* sailed into the Pacific Ocean and visited the Galápagos archipelago, a group of islands on the equator, far from land. Thanks to the governor, who pointed out that the giant tortoises indigenous to the archipelago are different from island to island, Darwin came to see that this variation held true of other fauna—the finches and mockingbirds in particular, which were peculiar to their specific island homes. This fact had to be important, especially since the various birds on the islands somewhat resembled birds on the South American mainland. On the continent Darwin had seen evidence of the same birds inhabiting sites from the steamy jungles of Brazil in the north to the snowy deserts of Patagonia in the south. But it was not until he returned to England that he understood the significance of these differences. Darwin showed his collection to the leading ornithologist of the day, John Gould, who determined that the Galápagos birds represented different species. With that information, Darwin made the big leap to evolution: the only way the differences could be explained was through *descent with modification.*

This became the central fact of evolution in Darwin's work. Darwin was never that focused on questions about the *path* of evolution (phylogeny)—on tracing which specific animal or plant groups descended from which other groups. After the Galápagos experience, he always thought of the path of evolution as being tree-like, with an original stem that branched up to the present. What was of greater concern to Darwin was the cause or causes of evolution. Without causes, he was no more than one among many evolutionists. With causes, he might become the Newton of biology.

As a graduate of the University of Cambridge (Newton's alma mater) and a friend and protégé of Whewell, Darwin knew full well that the best kind of science is that which has a cause at its heart, one which ties all together in a unified whole. For best results, this cause would be a kind of force, analogous to the force of gravity that, according to Newton, operated throughout the universe and was the key to all motions, terrestrial and celestial. Darwin wanted to find a biological force that would explain the evolution of life, one that operated throughout the living world and was the key to all organic motions. To this end Darwin worked feverishly for the next eighteen months—speculating and doing an extensive literature review. Quickly he realized that the probable key to change lies in some mechanism similar to the way in which animal and plant breeders effect change, namely, the selection and breeding of desirable offspring and the rejection of all others. Coming from the rural British Midlands, Darwin knew much about good husbandry. And now that so few people had to produce so much food for so many, thanks to the urbanization that accompanied industrialization, selection for improved quality of animals and plants was a more vital skill than ever.

But how was one to find a natural equivalent to the breeders' artificial selection? At the end of September 1838, Darwin read a conservative tract in political economy, *An Essay on the Principle of Population* (sixth edition, 1826) by the Reverend Thomas Robert Malthus. In this work, Malthus (another ordained Anglican clergyman) argued that potential population growth will always outstrip food supplies. Hence, unless people exercise "prudential restraint," ongoing struggles for existence are inevitable as people compete for the limited available food and other resources, including living space. This doctrine was very congenial to the social set in which Darwin lived. Malthus was arguing that state welfare only compounds the problems of the poor, because it leaves them with no incentive to do anything other than breed. Hence, people should be forced into working for a wage, and if the wage is low, well, so be it. Factory owners like the Wedgwoods loved this sort of thinking.

Darwin, while accepting the social sentiment (it comes out in later writings), turned Malthus's reasoning on its head. He pointed out that, in the animal and plant world, prudential restraint cannot exist, and since population growth certainly is not restricted to humankind, an ongoing struggle for survival inevitably occurs in the organic world. Success and failure in the struggle will (on average) be a function of the different characteristics possessed by organisms. Hence, the selection of better adapted organisms through competition in the natural world (that is, natural selection) will be ongoing, with the winners passing on their adaptive features to the next generation and the losers dying off. Over time, as environments alter, this selection will lead to full-blown organic change. "One may say there is a force like [a] hundred thousand wedges trying [to] force every kind of adapted structure into the gaps in the economy of Nature, or rather forming gaps by thrusting out weaker ones. The final cause of all this wedging, must be to sort out proper structure & adapt it to change" (Darwin Notebook D 135).

An idea is not a theory. Darwin worked hard to put things in a form that would be understood and accepted by professional scientists, writing up a 35-page "Sketch" of his thinking in 1842 and a 230-page *Essay* in 1844. At the same time, he cleaned up and published his geology and put out *Voyage of the Beagle,* a popular version of his official account of the time spent going around the globe—a travelogue that gained Darwin a wide and appreciative public audience. And he sat on his evolutionary ideas for fifteen years, turning in the meantime to his massive study of barnacles. Why Darwin delayed is not entirely clear. By this time, he had fallen sick with a mysterious ailment that was to plague him the rest of his days, and so undoubtedly he did not relish the prospect of a huge debate, which his ideas were bound to precipitate. Chambers's *Vestiges* had come out in 1844 and sparked a conflagration of controversy, with ferocious refutations by the likes of Sedgwick and Whewell, the very men who had led Darwin into a life of science. Darwin would hardly have welcomed a similar reception for his own ideas. So illness combined with an avoidance of conflict were surely factors in Darwin's tardiness.

But in the final analysis, perhaps there was no real reason; things may have just dragged on, until the day Darwin was forced into action. In 1858 he received from an unknown collector in the Malay peninsula—the Welsh-born Alfred Russel Wallace—a short essay containing virtually the same premises and conclusions that Darwin had discovered some twenty years earlier. Quickly, he wrote up his ideas yet again, and so the *Origin of Species* appeared in the fall of 1859.

The Origin of Species

The most significant fact about Darwin's world-shattering publication is the extent to which he had the basic outlines of the theory so early on, in his initial sketch. He begins with a discussion of artificial selection and the breeders' great successes. This strategy is partly heuristic, in that it allows readers to follow the path that Darwin himself took to discovery. But it is also partly justificatory, suggesting that what we humans can do, nature can surely do even better. After the breeders, Darwin discusses the variations among offspring that occur continuously in the wild—the critical building blocks of change. With the power of selective breeding and the fact of natural variation firmly in place, Darwin was now ready to introduce the struggle for existence and to follow this with a mechanism for change: natural selection (Darwin 1859, 63):

> A struggle for existence inevitably follows from the high rate at which all organic beings tend to increase. Every being, which during its natural lifetime produces several eggs or seeds, must suffer destruction during some period of its life, and during some season or occasional year, otherwise, on the principle of geometrical increase, its numbers would quickly become so inordinately great that no country could support the product. Hence, as more individuals are produced than can possibly survive, there must in every case be a struggle for existence, either one individual with another of the same species, or with the individuals of distinct species, or with the physical conditions of life. It is the doctrine of Malthus applied with manifold force to the whole animal and vegetable kingdoms; for in this case there can be no artificial increase of food, and no prudential restraint from marriage.

Note that, even more than a struggle for survival, Darwin needed a struggle for reproduction. It is no good (from an evolutionary point of view) to have the physique of Tarzan if you have the sexual drive of a philosopher. But with the struggle understood as a competition for survival *and* reproduction, and given that variation concurs continuously, natural selection follows at once (pp. 80–81):

> Let it be borne in mind in what an endless number of strange peculiarities our domestic productions, and, in a lesser degree, those under nature, vary; and how strong the hereditary tendency is. Under domestication, it may be truly said that the whole organization becomes in some degree plastic. Let it be borne in mind how infinitely complex and close-fitting are the mutual relations of all organic beings to each other and to their physical conditions of life. Can it, then, be thought improbable, seeing that variations useful to man have undoubtedly occurred, that other variations useful in some way to each being in the great and complex battle of life, should sometimes occur in the course of thousands of generations? If such do occur, can we doubt (remembering that many more individuals are born than can possibly survive) that individuals having any advantage, however slight, over others, would have the best chance of surviving and of procreating their

> kind? On the other hand we may feel sure that any variation in the least degree injurious would be rigidly destroyed. This preservation of favourable variations and the rejection of injurious variations, I call Natural Selection.

In the *Origin,* as well as in earlier versions of his theory, Darwin included a secondary selective mechanism—one that his grandfather had anticipated. Darwin argued that the struggle is not always for food and space; it is often a direct struggle for mates. This mechanism, *sexual selection,* Darwin divided into two parts: selection through male combat, as males compete with other males for access to females (the antlers of the deer would be a product of this); and selection through female choice, as males compete to attract the attention of females (the tail feathers of the peacock would be a product of this). I hardly need to stress that this two-pronged mechanism of sexual selection was inspired by the practices of breeders, who likewise select for male combative features (as in cocks and bulldogs) and for male beauty (as in the songs and feathers of prize birds).

Notice how sexual selection puts particular emphasis on competition between members of the same species. Selection is directed toward characteristics that benefit the individual rather than the species. Darwin tended always to see selection (including natural selection) as acting on the individual rather than the group—a view that would have been compatible with his middle-class industrialist background and certainly with the influence of Malthus, whose God deliberately set human against human in order to prod us to better our lot in life.

With the main mechanisms of change thus presented—there were various subsidiary mechanisms, including Lamarckism—Darwin introduced the famous metaphor of a tree. "The affinities of all the beings of the same class have sometimes been represented by a great tree. I believe this simile largely speaks the truth." The leaves and twigs at the top represent the species extant today. Then as we go down the branches, we have the great evolutionary paths of yesterday. All the way down we go until we reach the very first shared origins of life. "As buds give rise by growth to fresh buds, and these, if vigorous, branch out and overtop on all sides many a feebler branch, so by generation I believe it has been with the great Tree of Life, which fill with its dead and broken branches the crust of the earth, and covers the surface with its ever branching and beautiful ramifications" (Darwin 1859, 129–130).

And now, with some minor problems brushed away, Darwin was ready to present the second part of his theory. Here the influence of Whewell came out most fully. The older man, in praising Newton, had argued that the real triumph of the *Principia* was to show how so many different branches of physics could be brought together and united under one single causal mechanism. This strategy Whewell labeled a "consilience of inductions," and Darwin internalized and followed this principle with great care. For a good two-thirds of the *Origin,* Darwin takes us through the various branches of biological science—instinct, paleontology, biogeography, classification, embryology, morphology—showing that phenomena in these branches are explained by evolution through natural selection and that, conversely, these various branches point to and support the mechanism of evolution through selection.

Some problems arose, nevertheless. Social behavior—particularly that shown by ants, bees, and wasps—would be easily explicable if selection favored the group rather than the individual. From an individualistic perspective, why would sterile workers who devote their lives to the well-being of others in the nest or hive ever evolve? Darwin decided that in the case of the social insects (the hymenoptera), group members are so well integrated that it is permissible to treat the whole hive as a kind of super-organism, with the individual insects as parts of the whole. Just as selection can produce the eye, which exists for the benefit of the whole organism, so it can produce the worker ant or bee, which exists for the benefit of the whole hive.

Most other areas of biology fell into place with more ease. Geographical distribution (biogeography) was a triumph, as Darwin explained just why one finds various patterns of animal and plant distribution around the globe. What, for instance, accounts for the strange pattern of the Galápagos and other island groups? The explanation is simply that the founders of these isolated island denizens came by chance from the mainland and, once established, started to evolve and diversify under the new selective pressures to which they were now subject.

Embryology likewise was a particular point of pride for Darwin. Why, he asks, are embryos of different species sometimes very similar (man and dog, for instance), whereas the adults are very different? Darwin argued that selective forces operating at the embryological stage of development—in the womb—are more uniform and therefore produce similar outcomes, whereas selective forces operating at the adult stage of development—in the world—are highly diverse and produce very different outcomes. Here as always through his discussions of evolution Darwin turned to analogy with the world of the breeders in order to clarify and support the point at hand. "Fanciers select their horses, dogs, and pigeons, for breeding, when they are nearly grown up: they are indifferent whether the desired qualities and structures have been acquired earlier or later in life, if the full-grown animal possesses them" (Darwin 1859, 446). These claims were backed by experimental evidence, for Darwin had measured the young of pigeons and dogs and other specimens and found (even though breeders often denied this) that the differences were far less than those exhibited when the animals were full grown.

So we are led to the concluding passages of the *Origin*. Darwin never concealed that God was in the background of his work, and there are eight unself-conscious references to the Creator. But to a Creator of deist ilk. "Authors of the highest eminence seem to be fully satisfied with the view that each species has been independently created." This was not Darwin's position. "To my mind it accords better with what we know of the laws impressed on matter by the Creator, that the production and extinction of the past and present inhabitants of the world should have been due to secondary causes, like those determining the birth and death of the individual" (Darwin 1859, 488).

The Monkey Question

Of all the issues raised by the *Origin*, the "monkey question" is the one that absorbed the attention of Darwin's fellow-Victorians and caused them great doubt. If evolution is true, then are we humans little more than overgrown primates? Are we descended from the apes, rather than created in God's image? People on both sides of the Darwinian revolution worried incessantly about the status of humans. Sedgwick, a lifelong opponent of evolution, harped on the subject until his death. Lyell, although in a position to move to evolutionism well before Darwin, could never truly get God out of his science because he wanted to retain special status for humans. Thomas Huxley, as committed an evolutionist and materialist as there ever was, forever picked away at the scab of *Homo sapiens,* from his *Man's Place in Nature* (1863) right through to his final essay "Evolution and Ethics" (1893). How do you reconcile human nature with blind forces? Unlike his contemporaries, Darwin, from the first, was stone-cold certain that human beings are products of evolution, and that selection is the answer.

The first private notebook jottings that we have about the mechanism of selection address the subject of human intelligence. "An habitual action must some way affect the brain in a manner which can be transmitted.—this is analogous to a blacksmith having children with strong arms.—The other principle of those children. which *chance?* produced with strong arms, outliving the weaker one, may be applicable to the formation of instincts, independently of habits" (Darwin Notebook N, 42).

There were various factors that led to Darwin's hardline position, but most powerful of all were his experiences aboard *HMS Beagle.* Darwin had seen first-hand the savages down at the tip of South America. The Tierra del Fuegans were primitive beyond belief, yet they were human. Given our species' capacity for such uncivilized behavior, how could anyone think that we are all that different from the apes? What made Darwin's experience all the more striking was that the *Beagle* was returning three natives who had spent some time in England. They had become truly civilized, and yet after a week or two back on home turf they had reverted to their native ways. From Darwin's perspective, the line between the man of culture and learning and the brute was thin indeed. This lesson was one the ship's naturalist never forgot.

In the *Origin,* however, Darwin made a tactical decision not to play up the human question. He did not try to conceal his thinking, but he also did not want to let it swamp his general points about the process of evolution. So, in one of the great understatements of all time, he wrote: "Light will be thrown on the origin of man and his history" (Darwin 1859, 488). Darwin fully intended to get back to the topic and deal with it at length, as part of a grand writing plan to take up each section of the *Origin* and delve into it in much greater detail. But after the *Origin* was published, Darwin became more and more reluctant to commit himself to such an all-consuming program. Much more delicious and immediate projects competed for his attention, starting with a spritely little book on orchids and their fertilization that appeared in the early 1860s. But the human problem just would not go away, and finally in 1871 Darwin offered a major two-volume work, *The Descent of Man.*

This is a peculiar book. Most of it is not about our species at all but is an extended essay on sexual selection. The very peculiarity of this topic, however, provides the clue to its genesis. Although Darwin's personal relationship with A. R. Wallace was always good, the two men differed on many aspects of their jointly discovered child. As a teenager, Wallace had heard and been much impressed by the Scottish millowner and early socialist Robert Owen. This set up a life-long propensity to favor group thinking over individualism, and during the 1860s Darwin and Wallace had an extended correspondence about the possibility of a group-favoring form of selection, as opposed to selection that acts purely on the individual organism. Eventually they agreed to disagree, with Darwin opting for the individual and Wallace for the group.

More serious was a disagreement over our own species. When Wallace first put down on paper his thoughts about natural selection, he shared Darwin's naturalistic perspective on human origins, even penning a stimulating little essay on how we might have evolved. But then he became enamored with spiritualism and convinced that humankind could not have come into being through some random natural event. There must have been a Mind guiding our genesis. And to this conclusion, Wallace argued that many human characteristics could not have been the product of natural selection. He gave as examples hairlessness and great intelligence. Wallace, who had actually lived with natives, argued that a vast amount of unused brain power exists in primitive people; since the environment did not require or favor its development, this intelligence simply could not have been favored by selection.

Such a conclusion was anathema to Darwin, but he felt that Wallace had good arguments. Natural selection, unaided, could not produce many distinctive human features. But *sexual* selection—another topic on which Darwin and Wallace disagreed—could. Darwin argued that many of the characteristics that distinguish humans from other species came about through varying standards of beauty and desirability. The features mentioned by Wallace, although distinctive, do not take matters out of the range of evolution—and certainly not out of the range of science. And so we find a long discussion of sexual selection in general and then its application to humankind.

Why are men big, strong, and intelligent, Darwin asks? Because those who were got the best women. Why are women soft and yielding and domestic? Because those who were got the best men and did the best job with the family. Why are we hairless? Because that is what our ancestors found sexually exciting. Why (here he gets more specific) do Hottentot women have such big backsides (see p. 89)? Because this was the standard of beauty of the tribe, and the bravest braves got first pick.

As one might expect, all of this theorizing is wrapped up with some fairly standard Victorian sentiments about women and non-Anglo races. "It is generally admitted that with woman the powers of intuition, of rapid perception, and perhaps of imitation, are more strongly marked than in man; but some, at least, of these faculties are characteristic of the lower races, and therefore of a past and lower state of civilisation." If proof be needed, we can turn to social science, where we can infer that "if men are capable of decided eminence over women in many subjects, the average standard of mental power in man must be above that of woman" (Darwin 1871, 2.326–327).

By the 1870s Darwin's thinking was starting to settle into some fairly conventional modes. He may have been a revolutionary as a young man, but he was not always that far ahead of his tribe. There is little wonder, then, that by the time he died Darwin had became a major source of national pride. He was buried, by universal acclaim, right next to Isaac Newton in that English Valhalla, Westminster Abbey.

CHAPTER 6

A SUBJECT TOO PROFOUND

GENIUS THOUGH HE WAS, and father of a revolution, Darwin was a man of his time. He was incredibly lucky, being at the crossroads of so many influences—Darwins and Wedgwoods, Edinburgh and Cambridge, theism and deism, progress and Providence—and he made the most of his chances. But the parts came from the past—a past that he transformed and made into the future. This is true of his science, his social milieu, and above all his religion.

Whatever the stage of his thinking—theist, deist, or agnostic—Darwin framed all of his work in the context of the Judeo-Christian faith. For the ancient Greeks, for Plato or Aristotle, questions about ultimate origins were essentially meaningless. The world is eternal, even though the Demiurge may have done some shaping on matter already preexistent. Ultimately this world is one of endless change and decay, a building up and a falling down. It took the ancient Jews to make of our world a history, to make questions about ultimate origins meaningful, along with future expectations and consequences. Evolutionists, Charles Darwin in particular, are as much committed to the legitimacy and significance of this question as are believers in the literal truth of Genesis. In this sense, science versus religion, evolution versus Christianity, is a family quarrel—if indeed it is a quarrel at all.

Which brings us to the question of design and final causes. Given what we know about Darwin's personal background, we would expect, first, that he would have been exposed to teleological thinking—to an Anglicized, Platonic, external teleology in particular—and would have absorbed and accepted this thinking in some major respect. And second, we would expect Darwin to do something with it, transforming it to his own evolutionary ends. Both of these expectations prove true. Let us take them in turn, keeping always in mind the twin issues: the argument to complexity and the argument to design.

The Early Years

Darwin was not a professional philosopher or theologian, although he knew people who were—Whewell most obviously. But he was an educated Englishman, at a time when the classics were the major course in the educational diet. He knew about Plato and his theory of Forms or Ideas. He knew something about Aristotle and his general thinking about science. More significantly, he knew about natural theology, and specifically he knew about the arguments for the existence of God—the argument from design. Reading Paley's *Natural Theology* came with the territory, and Darwin responded very positively. Years later, in his *Autobiography,* he wrote that Paley's writings "gave me as much delight as did Euclid." In like spirit, just as the *Origin* was published, to a neighbor and fellow scientist Darwin wrote: "I do not think I ever admired a book more than Paley's Natural Theology: I could almost formerly have said it by heart" (letter to John Lubbock, November 22, 1859, in Darwin 1958, 59).

This way of thinking about the world would have been reinforced by Darwin's older friends and teachers at Cambridge. At the end of the 1820s, Cuvier's reputation was at its peak, and it was nigh tautological in Cambridge circles that the "conditions of existence"—the argument to complexity—was the key to understanding organic life. The key to a scientific understanding of organic life, that is. For instance, if we look at Henslow's approach to botany—whose lectures Darwin attended for all of his three years at Cambridge—we find that it was end-directed through and through. The "functions of vegetation" received detailed discussion. In the same vein, having discussed the propagation of plants by taking slips, Henslow wrote: "The greater number of plants, and *at least* all those which bear flowers, secure the continuation of their species by a distinct process, of a very different nature. This constitutes the function of 'reproduction,' properly so called; which consists in the formation of seeds, containing the germs of future individuals. This function of reproduction is to the species, what life is to the individual—a provision made for its continued duration on the earth" (Henslow 1836, p. 249).

The transcendental anatomy of Geoffroy and the *Naturphilosophen* was known in Britain at this time, but the Cambridge scientists, Sedgwick in particular, downplayed and discounted it. It was Paley's Platonic, externalist, final-cause view of life that Darwin imbibed and accepted. Nor, with respect to its implications for science, would this change much in the next decade. On the *Beagle* voyage, he was somewhat isolated from society and from new scientific currents. The exception was Lyell's geology, and in a way the exception proves the rule. By and large, the Cambridge group had little truck with uniformitarianism, and my suspicion is that it was precisely because Darwin was isolated—spending long nights alone with Lyell's attractive volumes—that he fell under the geologist's spell. Fortunately, although they differed from Lyell intellectually, members of the Cambridge group were on good personal terms with the author of the *Principles*, so when Darwin returned to England his new enthusiasm for Lyell's geology was no bar to his slipping right back into the old company and old ways.

Predictably, therefore, we find that even as Darwin was becoming an evolutionist and working toward and beyond natural selection, the end-directed way of thinking persisted. Indeed, the argument to complexity was absolutely central to his science. Very revealing are some private notebooks on the evolution question that he kept through the crucial months of discovery. Before the discovery of natural selection, we find Darwin worrying about sexuality and its purpose. Why are there two sexes, and what is their function? "I can scarcely doubt final cause is the adaptation of species to circumstance by principles, which I have given" (Darwin Notebook C, 236). Then right at the moment of discovery, the very point of the whole process: "One may say there is a force like hundred thousand wedges trying force into every kind of adapted structure into the gaps of in the oeconomy of Nature, or rather forming gaps by thrusting out weaker ones. The final cause of all this wedging, must be to sort out proper structure & adapt it to change" (D, 135e). And then, afterward, back to sexuality. Why is there sexuality? "My theory gives great final cause." And the answer being "otherwise, there would be as many species, as individuals" (E, 48).

As with Paley and Cuvier, this way of thinking was tied right in with adaptation or contrivance: organisms have characteristics like sexuality that serve certain ends. But this is precisely what Darwin's theory of natural selection was all about. Natural selection produces features like the eye and the hand, adaptations that exist in order to serve the ends of their possessors. In other words, Darwin took Paley's problem—How do we explain adaptation?—and treated it as a genuine problem, but he solved it through a lawbound process rather than through miraculous intervention. He shared Paley's conviction that the argument to complexity is real and that its conclusion is in need of solution. But he approached the argument to design in a way different from Paley, by taking the directly intervening God of miracles out of the scene. God was no longer in Darwin's science, or (to put it another way) God was no longer invoked as a direct supplement to the science. God may still be designing in the background, but he is now doing it at a distance, through the agency of law. The design-like nature of organisms was central for Darwin and in need of explanation, but not in a way that would bring God into science.

So, to recap in the language we have been using, Darwin accepted the argument to complexity: organic complexity was pervasive throughout the organic world and deserving of solution. He clearly interpreted this complexity as involving end-directed understanding—final cause. His goal, however, was to come up with a scientific explanation that could substitute for the argument to design, the argument to a creative intelligence. Darwin proceeded by breaking down the argument to design into a scientific part and a nonscientific part, giving an answer (natural selection) to the scientific part, and then saying that the nonscientific part is really not his concern as a scientist.

Nothing done by Darwin in the late 1830s committed him to denying a designer (although he certainly wanted to deny a miracle-intervening designer). He simply believed that one must give a naturalistic (that is, law-governed) scientific explanation for complexity. Questions about the connection between law and a designer remained, but these were not scientific questions. Which all means, incidentally, that we have to rethink the relation of humans to the rest of the world. We may be the superior species (Darwin always thought that) but not in a condescending theological fashion. Commenting on a report of Whewell's Bridgewater treatise, Darwin wrote that the author "quote[s] Whewell as profound, because he says length of days adapted to duration of sleep of man.!!! whole universe so adapted!!! & not man to Planets.—instance of arrogance!!" (D, 49).

Did Darwin think that absolutely everything is adapted, every last physical feature, no matter how useless or harmful it may be? Certainly not. In fact, as soon as he became an evolutionist, Darwin was looking at useless features as confirmation of his theory and as evidence to be used against a direct creationist position. He would expect rough edges, whereas a miracle-monger would not. In the margins of his copy of Whewell's *History of the Inductive Sciences,* which he read in 1837, Darwin scribbled sneers at Whewell's ebullient claims that adaptation is true of everything: "Mammae in Man" (3.456); "Shrivelled wings of those non-flying Coleoptera?!" (3.468); "When a man inherits a harelip or a diseased liver is this adaptation as much as Bullfinch to linseed.—doubtless it is in one sense, but not that in which these philosophers mean" (3.471, Di Gregorio and Gill 1990).

On his return to England, Darwin became quite close to Richard Owen, who would later gain a reputation as the arch-anti-Darwinian. But when he and Darwin were young men they enjoyed considerable social and intellectual intercourse. Owen, for instance, was given the task of examining and classifying some of the prize fossil specimens Darwin brought back from the *Beagle* voyage. They seem to have discussed the ideas of Geoffroy Saint-Hilaire a lot more easily and favorably than the older scientists in their group could have done. Through Owen, Darwin became sensitized to the facts of homology, realizing that this phenomenon was important and could not be dismissed as trivial. "Owen says relation of Osteology of birds to Reptiles shown in osteology of young Ostrich" (D, 35).

However, by now, and indeed forevermore, Darwin thought homology was a consequence of evolution—shared patterns that go back to a common ancestor—rather than something to be elevated to a principle or phenomenon in its own right. Homology needed an explanation, but it was not like adaptation in needing a causal explanation of its very own. This all comes through clearly in a discussion Darwin had with himself (in his notebooks) shortly after he became an evolutionist (1837). He was considering the Cuvier/Geoffroy conflict. He saw what was at issue, including the difficulties for the evolutionist: "The existence of plants, & their passage to animals appears greatest argument against theory of analogies" (B, 110). He saw also that Cuvier had to explain homology as a consequence of shared needs under the conditions of existence, and that this was stretching things beyond the power of the principle. In a passage quoted by Geoffroy (whom Darwin was reading), Cuvier wrote: "In one word, if under the unity of type one includes analogical resemblance . . . then this is only a principle subordinate to another which is higher and more fertile, namely that of the conditions of existence, of the harmony of the parts, of their coordination for the role which the animal will play in nature; this is the true philosophical principle where one discovers the possibilities of certain resemblances and the impossibility of certain others." Against this Darwin underlined "higher and more fertile" and wrote "I demur to this alone," adding "All this will follow from relation" and "The unity of course due to inheritance" (B, 112–115).

To the "Origin"

We move now to the first versions of Darwin's theory, the *Sketch* of 1842 and the *Essay* of 1844. Darwin's position on final cause was already there, along with all the other elements of his theory, and it was to remain virtually unchanged until Darwin laid down his pen at the end of his life. Natural selection (he now used this term) was the chief mechanism of evolutionary change, although for the first time we get, in the *Sketch*, explicit discussion of the secondary mechanism of sexual selection. Selection was not just a cause of change but, more significantly, of change in the direction of adaptive advantage. The analogy was with the changes brought about by breeders as they select for the features they want in animals and plants. Indeed, the analogy was underlined, for Darwin talked of nature as if it were a personified being, albeit much superior to humans: "As we assume his discrimination, and his forethought, and his steadiness of object, to be incomparably greater than those qualities in man, so we may suppose the beauty and complications of the adaptations of the new races and their differences from the original stock to be greater than in the domestic races produced by man's agency" (Darwin and Wallace 1958, 114–115).

Then with selection introduced, with the notion of adaptation—organized complexity, end-directed toward the survival and reproductive needs of the possessor—made central, Darwin was off to show how his theory could explain the organic world in a way far more efficient than any other, especially any other that did not suppose evolution. Almost at the end, Darwin raised the question of homology and demonstrated that it follows from the basic premises. "Unity of type" was important, but it was just one fact among many others that made more sense in the light of evolution. "Scarcely anything is more wonderful or has been oftener insisted on than that the organic beings in each great class, though living in the most distant climes and at periods immensely remote, though fitted to widely different ends in the economy of nature, yet all in their internal structure evince an obvious uniformity" (p. 220). The standard example given was the forelimb of the vertebrates, with similar bones in similar order but with very different functions.

Darwin was aware that the notion of homology had been extended from resemblances between different organisms to resemblances between parts of the same organism. Darwin accepted this, arguing that descent with modification can explain this as well. The paradigm example of such serial homology was the parts of the backbone of the vertebrates, whose top pieces supposedly were transformed into the bones of the skull. Darwin bought into this vertebrate theory of the skull, although later when Huxley demolished it, Darwin dropped it quietly.

The main point is that, when faced with such repetition of parts, the religious answer for Darwin was no answer at all. "These wonderful parts of the hoof, foot, hand, wing, paddle, both in living and extinct animals, being all constructed on the same framework, and again of the petals, stamina, germens, etc., being metamorphosed leaves, can by the creationist be viewed only as ultimate facts and incapable of explanation" (Darwin and Wallace 1958, 221–222). However, "On our theory of descent these facts [homologies] all necessarily follow for by this theory all the beings of any one class, say of the mammalia, are supposed to be descended from one parentstock, and to have been altered by such slight steps as man effects by the selection of chance domestic variations."

In the same mode as homology, Darwin introduced a discussion of "abortive or rudimentary organs," those features of the organic world that seem useless to the point of triviality or even danger. Certainly sometimes these organs get appropriated for other ends, as when "in the male of the marigold flower the pistil is abortive for its proper end of being impregnated, but serves to sweep the pollen out of the anthers ready to be borne by insects to the perfect pistils in other florets." But in other cases, as when teeth are found embedded in solid bone, "the boldest imagination will hardly venture to ascribe to them any function" (p. 235). From a Darwinian perspective, this sort of flotsam and jetsam was bound to be thrown up by the tides of evolution. "On the ordinary view of individual creations, I think that scarcely any class of facts in natural history are more wonderful or less capable of receiving explanation" (p. 236).

Finally, at the end of the discussion, Darwin made it very clear that, far from intending his arguments as a refutation of God's existence, he thought his position confirmed and supported it. "It accords with what we know of the law impressed on matter by the Creator, that the creation and extinction of forms, like the birth and death of individuals should be the effect of secondary [laws] means." Indeed, "It is derogatory that the Creator of countless systems of worlds should have created each of the myriads of creeping parasites and [slimy] worms which have swarmed each day of life on land and water on [this] one globe." God maximizes the good, even though this can only be done at the cost of some ill effects. In Darwin's case, we would expect God to create through law, and a cost and consequence of this would be pain and destruction—all for a greater good. "From death, famine, rapine, and the concealed war of nature we can see that the highest good, which we can conceive, the creation of the higher animals has directly come" (pp. 86–87). Although Darwin thought that, strictly speaking, none of these issues were scientific, he thought it politic to bring his thinking on these matters into his scientific discussion.

Between the early versions and the finished theory of the *Origin*, there was one subtle but significant change. In the *Sketch* and the *Essay*, Darwin saw adaptation as being perfect in the sense that the characteristics of organisms suited their needs precisely. By the time of the *Origin*, Darwin was thinking in a more relativistic mode. What counted was not perfection but being better than one's competitors. Take the eye, for example. In the *Essay*, he implied clearly that the eye is always perfect for its demands. Simple eyes were more than adequate for simple beings. "If the eye from its most complicated form can be shown to graduate into an exceedingly simple state—if selection can produce the smallest change, and if such a series exists, then it is clear (for in this work we have nothing to do with the first origin of organs in their simplest forms) that it may possibly have been acquired by gradual selection of slight, but in each case, useful deviations, and that each eye throughout the animal kingdom is not only most useful, but perfect for its possessor" (Darwin and Wallace 1958, 149). By the time of the *Origin*, Darwin's tune was changing. "Natural selection tends only to make each organic being as perfect as, or slightly more perfect than, the other inhabitants of the same country with which it has to struggle for existence. And we see that this is the standard of perfection attained under nature" (Darwin 1859, 201).

The eye now was an example of imperfection: "The correction for the aberration of light is said, on high authority, not to be perfect even in that most perfect organ, the human eye" (Darwin 1859, 202). By the sixth edition of the *Origin* (the final edition, appearing in 1872), Darwin quoted the German physiologist Hermann von Helmholtz on the eye's failings. "That which we have discovered in the way of inexactness and imperfection in the optical machine and in the image on the retina, is as nothing in comparison with the incongruities which we have just come across in the domain of the sensations. One might say the nature has taken delight in accumulating contradictions in order to remove all foundations from the theory of a pre-existing harmony between the external and internal worlds" (Darwin 1959, 374).

To counter criticisms of natural selection, in his revisions Darwin was trying to break the analogy between the eye and the telescope: "It is scarcely possible to avoid comparing the eye with a telescope. We know that this instrument has been perfected by the long-continued efforts of the highest human intellects; and we naturally infer that the eye has been formed by a somewhat analogous process. But may not this inference be presumptuous? Have we any right to assume that the Creator works by intellectual powers like those of man?" (Darwin 1859, 188). This argument may not have destroyed or invalidated the analogy between telescope and eye, but it certainly rendered it more complex than hitherto. Darwinian adaptation, Darwinian final cause, was relativized in a way that one would not expect of Paley's final causes.

Between the early versions of Darwin's theory and the *Origin* came his massive project on barnacles. It is a common grumble of morphologists, especially those concerned with classification, that adaptation is their greatest enemy. To see properly what relates to what, one must dig beneath the needs of the moment and find the underlying shared patterns, which an evolutionist would regard as evidence of shared ancestry. Darwin was already an evolutionist by the time he launched into the barnacle project, but at this point he was not yet ready to declare his hand. We therefore find him using Owen's concept of the archetype, trying first to work out the common archetype of the Crustacea and then of the barnacle (Cirripedia). The notion of homology was crucial to this project, as he compared one organism with another and then with yet a third, straining to see how all could be made to fit together into a coherent whole. "With regard to the homologies of these three pairs of limbs, my first impression was that they were the mandibles and the two pairs of maxillae in their earliest condition; but I consider this view as quite untenable, for several reasons; viz., the wide interval between their bases and the mouth itself,—the somewhat variable position of the mouth with respect to the legs,—and the position which the latter occupy in the second larval stage" (Darwin 1854b, 107). And again: "It is really beautiful to see how the homologies of the archetype cirripede, as deduced from the metamorphoses of other cirripedes, are plainly illustrated during the maturity of this degraded creature, and are demonstrated to be identical with those of the archetype Crustacean" (Darwin 1854b, 588).

But adaptation and final cause were far from forgotten, and Darwin did not hesitate to speculate on purposes and ends. One discovery that excited him was the existence of what he called "complemental males"—

barnacles which had degenerated so far that they were little more than swimming penises attached to sacks of sperm. But the reason for their existence altogether defeated him. "If the final cause of the existence of these Complemental Males be asked, no certain answer can be given; the vesiculae seminales in the hermaphrodite of Ibla quadrivalvis, appeared to be of small diameter; but on the other hand, the ova to be impregnated are fewer than in most Cirripedes" (Darwin 1854a, 214). "Certain answers" are not our worry. The point is that Darwin thought in terms of adaptation and final causes and considered it appropriate to ask such questions.

The Origin

By the end of the 1850s, the leading professional biologists of the day were thoroughly Germanicized. Men like Sedgwick and Whewell no longer had significant influence. Owen and (increasingly) Huxley were the forces in the scientific community. They used homology and archetype, or some equivalent, as the very basis of biological study. This meant, paradoxically, that in a way the science of the *Origin* already had a rather old-fashioned look at the time of its publication in 1859. Darwin was still obsessed with the preoccupations of the 1830s, and he insisted on giving answers in those terms. For him, what counted were causal answers—causal answers in a Newtonian mode—and the discussion had to be embedded in a context that acknowledged the natural theology of Paley and the comparative anatomy of Cuvier. Adaptation was still *the* question to be solved, as far as Darwin was concerned; homology was just a by-product of the operation of evolution. Function, purpose, and design continued to be paramount.

Darwin tied in natural selection with function: "With respect to the mammae of the higher animals, the most probable conjecture is that primordially the cutaneous glands over the whole surface of a marsupial sack secreted a nutritious fluid; and that these glands were improved in function through natural selection, and concentrated into a confined area, in which case they would have formed a mamma" (Darwin 1959, 263). Likewise, he spoke of purpose: "The common goose does not sift the water, but uses its beak exclusively for tearing or cutting herbage, for which purpose it is so well fitted, that it can crop grass closer than almost any other animal" (p. 249). And adaptation was, of course, a staple: "Can a more striking instance of adaptation be given than that of a woodpecker for climbing trees and seizing insects in the chinks of the bark?" (Darwin 1859, 184).

Final cause, as a significant rule of inquiry, had a key function. "With some animals the successive variations may have supervened at a very early period of life, or the steps may have been inherited at an earlier age than that at which they first occurred. In either of these cases, the young or embryo will closely resemble the mature parent-form, as we have seen with the short-faced tumbler" (Darwin 1959, 698). Why would this be so? "With respect to the final cause of the young in such groups not passing through any metamorphosis, we can see that this would follow from the following contingencies; namely, from the young having to provide at a very early age for their own wants, and from their following the same habits of life with their parents; for in this case, it would be indispensable for their existence that they should be modified in the same manner as their parents" (p. 698). On a related point, the "conditions of existence," or "conditions of life" as Darwin usually said, was one of the most-used concepts in the *Origin:* "Nor can organic beings, even if they were at any one time perfectly adapted to their conditions of life, have remained so, when their conditions changed, unless they themselves likewise changed; and no one will dispute that the physical conditions of each country, as well as the numbers and kinds of its inhabitants, have undergone many mutations" (Darwin 1959, 227).

Finally, unity of type—homology—and useless organs appeared in the *Origin* as they did in the earlier versions of Darwin's theory. They were significant in that they pointed to evolution and away from a directly designing and intervening god, but they too, like homology, were a by-product of the discussion rather than a starting point. Turning to comparative linguistics, an area of inquiry now rising in prominence, Darwin drew an obvious analogy: "Rudimentary organs may be compared with the letters in a word, still retained in the spelling, but become useless in the pronunciation, but which serve as a clue in seeking for its derivation. On the view of descent with modification, we may conclude that the existence of organs in a rudimentary, imperfect, and useless condition, or quite aborted, far from presenting a strange difficulty as they assuredly do, on the ordinary doctrine of creation, might even have been anticipated, and can be accounted for by the laws of inheritance" (Darwin 1859, 455–456).

Later Publications

In the years after the *Origin,* one of the most interesting of Darwin's publications was a little book on orchids that he published early in the 1860s. Unlike in some of the bigger books that would follow—*Variation under Domestication* and the *Descent of Man*—Darwin was not here dealing with unfinished business, answering questions provoked by the *Origin.* In the orchids book he laid out evolutionary biology as he hoped it would be done. "I think this little book will do good to the Origin, as it will show that I have worked hard at details, and it will perhaps, serve [to] illustrate how natural History may be worked under the belief of the modification of species" (letter to his publisher, John Murray, September 24, 1861). The title of the book, *On the Various Contrivances by which British and Foreign Orchids Are Fertilized by Insects, and on the Good Effects of Intercrossing* (1862), flags the fact that it is teleological through and through. Like Paley, Darwin was looking at the organic world as if it were an object of design: he was taking organized end-directed complexity as the absolutely crucial key to unlocking the secrets of the living world and its attributes. Contrivances are human objects created with an end in view, as in "I have invented a remarkable contrivance for getting the corks out of wine bottles." This was Darwin's perspective on the living world, just as it had been Paley's.

Once we get into the text of the orchids book, the argument to complexity was made over and over, with great effect. Thus right at the beginning, speaking of how an orchid is fertilized, Darwin described in detail the "complex mechanism" which causes this to happen. Little sacks of pollen are brushed by an insect as it pushes its way in, in search of nectar. But not just little sacks. Rather, little sacks (or balls) that are going to travel. "So viscid are these balls that whatever they touch they firmly stick to. Moreover the viscid matter has the peculiar chemical quality of setting, like a cement, hard and dry in a few minutes' time. As the anther-cells are open in front, when the insect withdraws its head . . . one pollinium, or both, will be withdrawn, firmly cemented to the object, projecting up like horns" (Darwin 1862, 15). And so on and so forth. Right through the entire book, the picture was one of complexity, adaptation, function, purpose. "When we consider the unusual and perfectly-adapted length, as well as the remarkable thinness, of the caudicles of the pollinia; when we see that the anther-cells naturally open, and that the masses of pollen, from their weight, slowly fall down to the exact level of the stigmatic surface, and are there made to vibrate to and fro by the slightest breath of wind till the stigma is struck; it is impossible to doubt that these points of structure and function, which occur in no other British Orchid, are specially adapted for self-fertilisation" (Darwin 1862, 65).

But even Darwin realized that the old natural theological perspective went only so far. He recognized that with evolution there is a significant shift sideways in our understanding of the causes of complexity. Unlike a human creator or a God who can design from scratch and call up tools and materials as needed, evolution through natural selection is constrained. It has to make do with what is at hand. It is rather like being stuck in the desert with a malfunctioning car. Unless you can get it going with the parts you have, you will perish. You cannot send to the garage for spares (Darwin 1862, 348):

> Although an organ may not have been originally formed for some special purpose, if it now serves for this end we are justified in saying that it is specially contrived for it. On the same principle, if a man were to make a machine for some special purpose, but were to use old wheels, springs, and pulleys, only slightly altered, the whole machine, with all its parts, might be said to be specially contrived for that purpose. Thus throughout nature almost every part of each living being has probably served, in a slightly modified condition, for diverse purposes, and has acted in the living machinery of many ancient and distinct specific forms.

Yet, constrained though it may be, the organism-as-if-it-were-designed-by-God picture was absolutely central to Darwin's thinking in 1862, as it always had been—a point that made even some of Darwin's closest supporters uncomfortable. Wallace, for instance, wrote to Darwin about the term "natural selection," worrying that it was too anthropomorphic. It led, he said, to Darwin's "frequently personifying nature as 'selecting,' as 'preferring,' as 'seeking only the good of the species,' etc., etc." (Darwin and Seward 1903, 1.267–268). He urged Darwin to accept Herbert Spencer's alternative term "survival of the fittest" because it had fewer teleological connotations and was less likely to be "so much misrepresented and misunderstood" (p. 269). Darwin did in fact introduce this term into later editions of the *Origin*. He also regretted that his youthful enthusiasm for Paley and the argument from design had led him to too great a conviction that adaptation rules supreme. But really these adjustments were just cosmetic. In his heart, Darwin seems never to have wavered, and he responded to those who criticized the term "selection" by pointing out that it was a metaphor, and who can do science without being metaphorical? "No one objects to chemists speaking of 'elective affinity;' and certainly an acid has no more choice in combining with a base, than the conditions of life have in determining whether or not a new form be selected or preserved" (Darwin 1868, 1.6).

By the time Darwin finished his later works on our own species, *The Descent of Man* (1871) and *The Expression of the Emotions in Man and Animals* (1872), Darwin's backsliding days were over. Because of Wallace's own apostasy on the question of human origins, sexual selection started now to play a major role in Darwin's thinking, but this mechanism is as much end-directed as is natural selection. Remember those Hottentot bottoms: what were they for, if not to attract the males of the group? They were body parts with a purpose. It was the same with mental features. "The half-human progenitors of man, and of men in a savage state, have struggled together through many generations for the possession of the females. But mere bodily strength and size would do little for victory, unless associated with courage, perseverance, and determined energy" (Darwin 1871, 2.327). The younger men are always competing between themselves and trying to grab the goodies of the older men. Combine these trials with outside forces, the need to find food and to fend off predators and enemies, and you need lashings of "observation, reason, imagination or invention." The consequence is that "these various faculties will thus have continually been put to the test, and selected during manhood" (2.327–328).

Culture too got roped into the act by Darwin. Some aspects of what humans have wrought he judged to be deleterious, backfiring from an evolutionary perspective. Ethics is at least partially cultural. But, biologically speaking, ethics seems not to be an entirely unmixed blessing. Modern medicine, an outgrowth of the ethical value we place on the lives of our fellow human beings, seems to preserve the weakest in the group. "With savages, the weak in body or mind are soon eliminated; and those that survive commonly exhibit a vigorous state of health. We civilised men, on the other hand, do our utmost to check the process of elimination; we build asylums for the imbecile, the maimed, and the sick; we institute poor laws; and our medical men exert their utmost skill to save the life of every one to the last moment" (Darwin 1871, 1.168). I doubt, though, that Darwin was ever seriously worried about selection being a moral problem. The God of Robert Malthus was no tender-hearted do-gooder.

In any case, by this stage of his life, Darwin had moved on from deism to agnosticism. Theological worries about opposing selection in the name of a God-imposed morality were otiose to him. And this is quite apart from the fact that all is not entirely black. Sometimes culture works to good ends—no grandson of Josiah Wedgwood could be insensitive to the virtues of capitalism. It is true that we do not all start life with the same benefits, "but this is far from an unmixed evil; for without the accumulation of capital the arts could not progress; and it is chiefly through their power that the civilised races have extended, and are now everywhere, extending their range, so as to take the place of the lower races" (1.169).

Darwin and Design

Let us take a moment to tie things together. We have seen that Darwin became an evolutionist as much because of his religious beliefs as despite them. He had moved from the miracle-laden, theistic Christianity of his youth to a law-stressing deism, but there is no reason to think that—right through the formulation of his theory and on to the writing of the *Origin*—this belief was half-hearted or insincere. He stressed that his conception of the Creator was superior to that of the naive theist, and he meant this. Theologically speaking, creation through law was superior to creation through miracle, and pragmatically or scientifically speaking it explained much that was otherwise inexplicable—as well as letting God off the hook for much that is distasteful about the world of life.

However, overlaid on Darwin's evolutionism was his urge to find a cause for evolution—a force of the kind posited by Newton, two centuries before. Here, Darwin's Christian training came to the fore. Malthus made sense because Englishmen of Darwin's day expected God to have designed things so as to spur us to action rather than allowing us to rest idle and content with our lot. The struggle for existence might be a stern master, but without it nothing good would be accomplished. More important, Paley had pinpointed the crucial distinguishing feature of the organic world—a feature reemphasized by the great Cuvier—namely, its design-like nature: its contrivance—what we have been speaking of as end-directed, organized complexity. Darwin always accepted that an explanation for this complexity, an explanation for the adapted nature of organisms, was the key question for the evolutionist. Darwin accepted the argument to complexity, which saw an organic world full of things demanding explanation in terms of final causes, and he then set about to give his answer to the puzzle that Paley and others had explained in terms of a creative designer. But for Darwin, despite the theology in his background, his immediate solution had to be a scientific, not a theological, one.

It was to this organized complexity—contrivance, adaptation—that Darwin's concept of natural selection spoke successfully. When he eventually came to see that he had shifted the problem sideways somewhat, he simply adjusted his position to be more relativistic and less perfectionist. Selection, in his revised view, cares only about winning, not about perfection. And selection can work only with the materials already in existence. So, in a sense, as Darwin realized, he was producing something of a Heath Robinson perspective on the natural world, where solutions are cobbled together from pieces at hand and ends are not necessarily achieved in the most rational means possible (see page 107). But overall, it was the organic, organized complexity question that absorbed Darwin's attention, and it was the artifact metaphor as mediated by natural selection that gave him his solution.

And the nice thing is that, as in the best kind of scientific advance, the anomalies and tensions became supporting pieces of information. For Ptolemy, the inferior/superior planetary distinction was simply an unwanted and unexplained phenomenon; for Copernicus, it nicely confirmed that some planets were closer to the sun than the earth, and others farther. For the pre-Darwinian teleologists—Cuvier and Whewell especially—vestigial organs were irritating and homologies downright threatening; for the pre-Darwinian nonfunctionalists—people like the *Naturphilosophen*—homology was all-important, but the reason for its existence had to be plunged into nonscientific mystery, often nonscientific Platonic mystery. For Darwin, on the other hand, vestigial organs come from the string-and-sealing-wax way in which selection works. Selection is not the kind of cause that produces perfection—it works with what it has, and as a result some organisms do better than others. Fortunately, doing better than others is good enough. For Darwin, homology—unity of type—is a fundamental confirmation of his theory.

Inasmuch as he was obsessed with the argument to complexity, Darwin painted a thoroughly teleological world picture. He took up well-worn themes and problems, but his solution was fresh and innovative. The many diverse reactions to this theory are hardly a surprise, for he blended old and new into creative, unexpected patterns. He faced a Platonic question and offered an Aristotelian answer. That is, he showed how to get purpose without directly invoking a designer—natural selection gets things done according to blind law without making direct mention of mind. The teleology is internal.

But finally, overall, one might say that Darwin's biology, more than being Aristotelian, was Kantian. He has no place for forces that lie beyond physics and chemistry. Yet at the same time he recognized a dimension to biological understanding that transcends physics and chemistry. There are meaningful questions and answers which are not those of physics and chemistry. Because of its design-like nature, the living world asks more of these questions than the nonliving world.

As a Christian young man, Darwin clearly viewed the world—especially the organic world—as impregnated with values that in some absolute sense are truly good, and very much to the benefit of our own species. We are the special creation, made in his image. My suspicion is that, within Darwin's inner mind, this world picture never darkened entirely. To the end of his days, Darwin saw evolution as leading to humankind, which, for a good Victorian, is no bad thing. But as a professional scientist, he saw that the introduction of any kind of absolute value into science is inappropriate. One may hold values, but they are not a proper part of the scientific enterprise.

Finally, let us get back explicitly to the argument to design, which has been rather neglected in our discussion, and legitimately so, given that Darwin's great contribution was to science and not to theology. Right through the writing and publishing of the *Origin,* Darwin held on to some positive version of the theological argument, but his God stood back from his creation and had absolutely no direct role or position to play in science. At this point, Humean criticisms came right back into play, finally becoming strong enough to tip Darwin himself over into agnosticism. We know that he was much perturbed by the pain and pointlessness of the early death of his favorite child, Annie. Although not yet definitive, these worries had started to kick in by 1860, in a letter to Asa Gray of May 22:

> With respect to the theological view of the question; this is always painful to me.—I am bewildered.—I had no intention to write atheistically. But I own that I cannot see, as plainly as others do, & as I sh^{d}. wish to do, evidence of design & beneficence on all sides of us. There seems to me too much misery in the world. I cannot persuade myself that a beneficent & omnipotent God would have designedly created the Ichneumonidae with the express intention of their feeding within the living bodies of caterpillars, or that a cat should play with mice. Not believing this, I see no necessity in the belief that the eye was expressly designed. On the other hand I cannot anyhow be contented to view this wonderful universe & especially the nature of man, & to conclude that everything is the result of brute force. I am inclined to look at everything as resulting from designed laws, with the details, whether good or bad, left to the working out of what we may call chance. Not that this notion *at all* satisfies me. I feel most deeply that the whole subject is too profound for the human intellect. A dog might as well speculate on the mind of Newton.—Let each man hope & believe what he can.

Later, although apparently he always had flashes of conviction that a designer does exist, Darwin went even farther down the track of disbelief. We might say that he planted a bomb under Victorian teleology; whether it exploded with such force as to destroy absolutely any possibility of a theological argument to design is a matter yet to be discussed. Certainly the old argument to design could not remain as it was. There was now a bomb, and it must be defused else it would detonate. And thus we can see why so much controversy swirled about Darwin and the whole question of end-directedness. Both sides were right: those who claimed that Darwinian evolution reaffirmed purpose in the living world and those who claimed that Darwin forever destroyed it. It is just that the two sides were offering answers to different questions. Yes, Darwin endorsed the argument to complexity; but he changed the argument to design forever.

CHAPTER 7

DARWINIAN AGAINST DARWINIAN

WITHIN A VERY SHORT TIME after the *Origin,* evolution as fact had gained wide acceptance, especially among professional scientists. The case that Darwin built for shared origins was just too strong to be denied—not that most people by this time wanted to deny it anyway. If God wished to create through unbroken law, then that was God's business, not ours. Darwin of course wanted more. He hoped to found a functioning professional science of evolutionary studies, empirical and experimental, that would have natural selection as its causal core—something very much along the lines of astronomy, with Newton's gravitational force at its center, or (a mid-Victorian favorite) optics, with the wave theory of light at its center. But this was not to be. A professional evolutionism did emerge, but it was not Darwinian in any real sense.

In fact, it was not primarily English, being located essentially in German universities and owing more to *Naturphilosoph* roots than to anything coming from the *Origin.* Spearheaded by the German biologist Ernst Haeckel (1834–1919), this "evolutionary morphology" was little concerned with causes. It was obsessed, rather, with tracing relationships among organisms and above all teasing out their evolutionary histories—their phylogenies. Focusing on homologies and finding embryological analogies everywhere (it was Haeckel who promoted the "biogenetic law" that "ontogeny recapitulates phylogeny," an updated but today outdated version of the traditional analogy between individual and group development), it spun ever-more fantastical scenarios about the paths of life. Its flourishing progressionist evolutionary tree was quite the equal of anything that existed in the Garden of Eden.

Evolutionary morphology proved not to be a very successful professional science, in large part because of the many gaps in the fossil record it was trying to interpret. These gaps led to wild speculations and outright contradictions that undermined the field. Ironically, during this time the fossil record was being revealed as never before. Archaeopteryx—part reptile, part bird, the archetypal "missing link"—appeared at the beginning of the 1860s; Neanderthal man was around, giving rise to many happy speculations about which Celtic outposts still house his descendants; and after 1870 the fabulous dinosaur and mammalian deposits of the American West became available for study. Yet, despite these riches, bright young biologists moved to more inviting studies, such as cellular biology and—after the rediscovery in 1900 of Mendel's work—the newly developing science of heredity or genetics. William Bateson, the father of British Mendelian genetics, was open about the failures of evolutionary morphology. "Discussion of evolution came to an end because it was obvious that no progress was being made. Morphology having been explored in its minutest corners, we turned elsewhere" (Bateson 1928, 390). Failing to make the grade as a cutting-edge, university-type discipline, evolutionary morphology found its true home in museums, which welcomed displays of past life forms, especially those wonderful fossil dinosaurs that were being sent back from the New World.

In another important respect, evolutionism was enjoying a grand success. This came about through the work above all of "Darwin's bulldog," Thomas Henry Huxley. In the 1860s Huxley and his friends were at the front of a drive to reform British science, industry, government, and civil society (with effects in other countries, including America). For all that Britain was now building the largest empire the world had ever known, it was still a land sadly unfitted for the industrialized, urban, secular society that it had become. The first sign of trouble was the disastrous showing in the Crimean War (1854–1856), followed by the Indian mutiny (1857). The British medical profession, surrounded by quacks and pretenders, was failing; huge cities were in need of sanitation, housing, and transportation; the population was largely uneducated; and the civil service operated according to rules of patronage rather than merit. Progressive men like Huxley (and some women, such as Florence Nightingale) set about to change all this—to drag Britain out of the eighteenth century and turn its attention to the demands of the twentieth.

Huxley's own field of interest and influence was science, especially science education. He became an administrator at the newly founded Science University in South Kensington, where he reformed the curriculum, established examinations, networked with the government and other people with power and funds, and much more. As an administrator, Huxley understood that simply founding a scientific discipline would not be enough to ensure its success. Above all, he had to find jobs for his students. Although he was no physiologist, Huxley recognized this as a field he could sell to the medical profession. Its leaders, who now felt a need to stop killing people and start curing them, were happy to listen to Huxley's pitch that training in physiology was the ideal background for new doctors. And the universities were happy to take up Huxley-sponsored candidates as instructors.

Huxley sold morphology, his own specialty, to the teaching profession. Dissecting a dog fish to learn comparative anatomy was far better training for the new Victorian, he said, than the study of outmoded classics by Plato and Aristotle, Horace and Virgil. "Sit down before fact as a little child, be prepared to give up every preconceived notion, follow humbly wherever and to whatever abysses nature leads, or you shall learn nothing" (Huxley 1900, 1.219). As a newly elected member of the London School Board, Huxley set up summer schools for the training of science educators and then sent his acolytes forth to preach the message far and wide. The novelist H. G. Wells, Huxley's most famous student, put all sorts of evolutionary ideas into his early novels, especially the *Time Machine,* with its tale of the human race in the far future.

So where did Huxley the ardent evolutionist fit Darwinism into all of this reform? He could not see a practical use for evolutionary studies, and he had little personal interest in making a causal science of the subject. Huxley could, however, see a social role for evolution, especially that phylogeny-tracing progressivism of the Germans and their followers. Christianity, particularly the established Church of England, was identified with the old order. It stood for the forces of reaction against change, and it held a vice-like grip on education. Huxley and his friends needed a new ideology for a new age—they needed their own religion or religion-equivalent to offer the public. What better choice than evolution? It became a kind of secular religion, a modern metaphysics, one that answered questions about origins, as did Christianity; that told us our place in the scheme of things, as did religion; that offered an overall meaning to history, namely, Progress rather than Providence.

Bluntly, Huxley wrote to a friend of "a course of instruction in Biology which I am giving to Schoolmasters—with the view of converting them into scientific missionaries to convert the Christian Heathen of these islands to the true faith" (Huxley 1900, 2.59–60). Admittedly a joke, but many a true word has been spoken in jest. Not for nothing was he known in the popular press as "Pope Huxley." Striving to achieve his ends, night after night Huxley preached the gospel of evolution: at workingmen's clubs, in public debates with clerics, at scientific and other associations, in journals and newspapers and magazines. And always it was a story with a message about how we came to be, who we are now, and what the course of history can and should become.

A good religion has guides for proper conduct—love your neighbor as yourself and that sort of thing—and accordingly Huxley and his friends found ethical guidelines in evolution. For all that Darwin's *Descent* contained some unambiguous moral sentiments about the need to cherish the white race and promote capitalism, such moralizing was not really Darwin's intent. But his fellow English evolutionist Herbert Spencer felt no such qualms. In publication after publication, he proselytized for what became known (somewhat misleadingly) as "Social Darwinism." Only by following the rules of evolution, argued Spencer, can one hope to maintain and achieve yet further progress, social and biological. Happily crossing the barrier between the way things are to the way things ought to be, Spencer and his many admirers—as much in America and the rest of the world as in his English homeland—argued that one must promote the ways of evolution in the name of the higher good.

Hence, evolution after Darwin became a vehicle for moral and social messages. It was drenched with value-talk of the most absolute and positive kind, while failing as a causal and quantifiable, experimental and empirical science. From a professional standpoint, evolution was at most a second-rate enterprise: too much missing between premises and conclusions, and an inadequate methodology. But as an ideology to replace a no-longer-sufficient Christianity, as a secular religion, it was a smashing success.

This is directly important to us, for the whole question of teleology—

final cause, design, function—is partly scientific (the argument to organized complexity) and partly philosophical/theological (the argument to design and a designer). If, after Darwin, evolution had succeeded as a purely scientific subject, then we would expect to find a great deal of discussion about the argument to design simply dropping out of sight. Or at the very least, we would expect any philosophical/theological discussion to be the province exclusively of philosophers and theologians. Scientists would have their own problems to solve, and whatever teleology they invoked would be restricted to these issues. The argument to design would become more or less irrelevant from the perspective of science, and hence of evolution.

However, since in fact evolution became both less than a science and more than science, all bets are off. Two ardent Darwinians who emerged in the decades after the *Origin* embodied the conflicts and alternate interpretations of this period. Huxley, our man in England, a nonbeliever in the Christian message, used evolution to further his secular agenda, while his nearest counterpart in America, Asa Gray, very much a believer in the Christian message, still found reasons to promote evolution's virtues.

Darwin's Bulldog, Thomas Henry Huxley

Whatever qualifications one might want to make, T. H. Huxley truly was the greatest Darwinian of them all. He came from the lowest segment of the British middle-class—his schoolteacher father slid in and out of madness—yet he pushed himself forward to prominence through talent and hard work. Backed by scholarships, he trained as a physician and then joined the Royal Navy as a surgeon, aboard *HMS Rattlesnake.* It was on a trip to the South Seas in the late 1840s that Huxley laid the foundations of his success as a scientist, for he performed delicate dissections of the fragile invertebrates that he was able to fish up from over the side of the ship. He admitted that he was not primarily interested in function (and would have had trouble studying it anyway). For him, the thrill was anatomy and the discerning of similarities or homologies. "I am afraid there is very little of the genuine naturalist in me. I never collected anything, and species work was always a burden to me; what I cared for was the architectural and engineering part of the business, the working out the wonderful unity of plan in the thousands and thousands of diverse living constructions, and the modifications of similar apparatuses to serve diverse ends" (Huxley 1900, 1.7–8).

Huxley's first major paper, one that was to pave his way into the scientific community, was on the Medusae (a family of jellyfish), explicitly their anatomy and relationships with other organisms. "In order to demonstrate that a real affinity exists among different classes of animals, it is not sufficient merely to point out that certain similarities and analogies exist among them; it must be shown that they are constructed upon the same anatomical type, that, in fact, their organs are homologous" (Huxley 1903, 1.23). This was Huxley's passion throughout his adult life. Back in England, he set about establishing a career as a professional biologist and finding work to support himself. After some difficulties, Huxley got a job at the Royal School of Mines. This gave him an income and a support for his morphological labors, and he continued in the same pattern of work as before.

Realizing how much morphology owes to German thought, Huxley, in typical fashion, set about teaching himself the language. And this endeavor led to an even greater and more explicit acknowledgment of the continental tradition. "The biological science of the last half-century is honourably distinguished from that of preceding epochs, by the constantly increasing prominence of the idea, that a community of plan is discernible amidst the manifold diversities of organic structure" (Huxley 1903, 1.538).

Such thinking was characteristic of England's leading morphologist, Richard Owen. Upon Huxley's return from his voyage, as he worked his way into the scientific community, his initial interactions with Owen were friendly. Soon, however, the two men fell out. Owen was thin-skinned and jealous of his own position; Huxley was thick-skinned and jealous of Owen's position. Huxley started to worry away at parts of Owen's work, ridiculing the results and throwing ill light on the investigator.

The real quarrel was not scientific but metaphysical. As a modern-day, forward-looking Victorian (the utilitarian John Stuart Mill was now the country's leading philosopher), Huxley was setting himself up as an empiricist rather than an idealist. He was prepared to use the language of the "archetype," but he did not want his work identified with some real Platonic Form, existing independently of sensed reality. In his critique of Owen, after describing many isomorphisms between the skulls of vertebrates, Huxley concluded that it is "needless to seek for further evidence of their unity of plan." Digging away at the older man: "There is no harm in calling such a convenient diagram the 'Archetype' of the skull, but I prefer to avoid a word whose connotation is so fundamentally opposed to the spirit of modern science" (Huxley 1903, 1.571).

Together with a shift from idealism to empiricism in the philosophical world, Huxley made the shift from Christianity to what he labeled "agnosticism." So, naturally, he did not want to have any truck with claims about Ideas being the thoughts of God or such things. "I cannot see one shadow or tittle of evidence that the great unknown underlying the phenomena of the universe stands to us in the relation of a Father—loves us and cares for us as Christianity asserts" (Huxley 1900, 1.260). The rule of impersonal law—what Huxley dubbed "scientific Calvinism"—was everything. Sometimes law works for you. Sometimes law works against you. "This seems to me to be the relation which exists between the cause of the phenomena of this universe and myself. I submit to it with implicit obedience and perfect cheerfulness, and the more because my small intelligence does not see how any other arrangement could possibly be got to work as the world is constituted."

At the beginning of the 1850s, Huxley had opposed evolution. Indeed, trying to establish his credentials as a proper professional biologist, he had written a review of Robert Chambers's *Vestiges* more critical and scornful than any except perhaps that of Sedgwick—an interesting exercise, given that Huxley was just coming out from under an enthusiasm for MacLeay's quinary system. By the end of the decade, Huxley had had his conversion experience and, like Saint Paul, was going to spend the rest of his life unceasingly spreading the good word. How did this metamorphosis come about? In addition to the influence of Herbert Spencer, also important was Huxley's growing friendship with Darwin and his circle—Lyell the geologist, Hooker the botanist, and others. This group served as psychological support to counter the hostility of Owen and his followers. And then there was Huxley's scientific Calvinism, backed by agnosticism. From denial of the gospel of evolution, he became its leading apostle. But how then was Huxley to deal with teleology, so vital to the Christian position with which he saw himself locked in mortal combat? It could not go unquestioned, nor did it.

Huxley on Function

Naturally enough, Huxley wanted nothing to do with one level of the final-causal argument, the argument to design and to a designer. As far as Huxley was concerned, Darwin had wiped that out completely, and that was the end of it. Certainly Darwin had wiped it out if you understand the argument to design as claiming that the only possible solution, once design has been accepted, is a conscious intelligence—a deity of some kind, Christian or other. By a wonderful stroke of good fortune, Huxley was given the *Origin* to review for the London *Times,* and on Boxing Day (December 26) 1859 Huxley offered England's leading minds and civil authorities a very favorable account of the Darwinian thesis. He raised the question of traditional teleological understanding only to discount it brutally. At the end of the review, Darwin was much praised for steering his countrymen away from teleology. "The path he bids us follow professes to be, not a mere airy track, fabricated of ideal cobwebs, but a solid and broad bridge of facts. If it be so, it will carry us safely over many a chasm in our knowledge, and lead us to a region free from the snares of those fascinating but barren virgins, the Final Causes, against whom a high authority has so justly warned us" (Huxley 1859, 21). One wonders what the squires in their manor houses and the parsons in their vicarages thought of that, as they tucked into the leftovers from the season's repast.

The full force of Huxley's negative views on teleology (the teleology that pitchforks one into accepting a designer) came out a year or two later, when he was responding to criticisms of Darwin's ideas, specifically to those of a German biologist who had accused Darwin of leaving too much teleology in his theorizing. "It is singular how differently one and the same book will impress different minds. That which struck the present writer most forcibly on his first perusal of the 'Origin of Species' was the conviction that Teleology, as commonly understood, had received its deathblow [at] Mr Darwin's hands" (Huxley 1864, 82). And then: "If we apprehend the spirit of the 'Origin of Species' rightly, then, nothing can be more entirely and absolutely opposed to Teleology, as it is commonly understood, than the Darwinian Theory. So far from being a 'Teleologist in the fullest sense of the word,' we should deny that he is a Teleologist in the ordinary sense at all" (p. 86).

To deny that the argument to design has force—a denial that was crucial to Huxley's strategy—you need good reason to think that the argument to organized complexity raises no problems that only a theological approach can solve. At one level, Huxley appreciated that providing these good reasons was what the *Origin*—with its mechanism of natural selection—set out precisely to do. Darwin showed that the organic world does not exhibit characteristics that can be answered only by invoking a designer. "According to Teleology, each organism is like a rifle bullet fired straight at a mark; according to Darwin, organisms are like grapeshot of which one hits something and the rest fall wide." Darwin therefore blocks or makes unnecessary the appeal to intelligence. "For the teleologist an organism exists because it was made for the conditions in which it is found; for the Darwinian an organism exists because, out of many of its kind, it is the only one which has been able to persist in the conditions in which it is found."

Crucially, Huxley appreciated that Darwinism is relativistic in a way that a perfect, designing intelligence would not be. "Teleology implies that the organs of every organism are perfect and cannot be improved; the Darwinian theory simply affirms that they work well enough to enable the organism to hold its own against such competitors as it has met with, but admits the possibility of indefinite improvement." Using the example of cats being well adapted for catching mice and birds, Huxley showed how Darwinism takes things from the opposite direction to that of the traditional teleologist. Cats are as they are, because (thanks to their qualities) they succeeded, rather than because they were designed (with their qualities) for success. "Far from imagining that cats exist *in order* to catch mice well, Darwinism supposes that cats exist *because* they catch mice well—mousing being not the end, but the condition, of their existence" (Huxley 1864, 84–86). That is why we have cats as we do, at the end of a long, slow evolutionary process. Not because someone (or Someone) sat down and sketched out the form of cats and then set about making them, miraculously.

However, there was another level to Huxley's thinking at this point—a level that was at least as important, if not more so. Basically, Huxley was just not that smitten with the argument to organized complexity. As far as he was concerned, adaptation was no big deal, and certainly not at the front of the modern biologist's consciousness. Huxley was a product of the late 1840s and 1850s, when German idealistic morphology was in full force. He may have given this theory a materialistic interpretation, but he was still pushing homologies, and form over function, nevertheless. Paley was not on Huxley's reading list, and Cuvier was outmoded—it was Huxley who pointed out that it was really comparative anatomy that guided Cuvier to his great successes, rather than the Frenchman's own fancy conditions-of-existence theorizing. As an anatomist first, and then, during the 1860s, increasingly as a paleontologist, Huxley was only truly excited about his subjects after they were dead, precisely when they were no longer able to test their adaptations against a harsh, unforgiving environment.

Adaptation was, if anything, a hindrance to the morphologist's aim of discerning relationships. Huxley accepted adaptation, but he was fundamentally indifferent to it. Darwin could never have invited his reader, as did Huxley just a year or two before the *Origin* appeared, to ask, with regard to birds and butterflies, if "their beauty of outline and colour . . . are any good to the animals? that they perform any of the actions of their lives more easily and better for being bright and graceful, rather than if they were dull and plain?" Just a few years before Darwin penned his little book on adaptation in orchids, Huxley, ironically, had this to say on the subject: "Who has ever dreamed of finding an utilitarian purpose in the forms and colours of flowers, in the sculpture of pollen-grains, in the varied figures of the fronds of ferns?" (Huxley 1903, 1.311).

Not much changed in Huxley's mind once the *Origin* was openly on the table. To Darwin's chagrin, six lectures that Huxley gave to "Working Men" in 1863 made virtually no mention of adaptation or of the significance of selection in achieving it. Even in reviews of the *Origin*, Huxley did little better. In the instant review that appeared in the *Times* at the end of 1859, although there was certainly mention of selection, Huxley himself pushed for "saltations"—nondirected jumps in the evolutionary process from one form to another (as fox into dog)—a concept that an adaptationist could never accept. In the early 1840s Darwin had explicitly and firmly barred saltations, precisely because they leap over the process of adaptation. The kinds of jumps we see in nature are drastically maladaptive, and this, Darwin pointed out, is the general way of nature. The bigger the random move, the more likely that things will not function properly. Appreciating this fact, Huxley in his review dwelt on features that are not adaptive—features Darwin recognized but hardly stressed as needing a major mechanism other than selection.

Huxley did not deny adaptation or even that teleologists had a point in thinking that something about biological thinking is distinctive. Writing of Darwin's achievement, he allowed: "We should say that, apart from his merits as a naturalist, he has rendered a most remarkable service to philosophical thought by enabling the student of Nature to recognise, to their fullest extent, those adaptations to purpose which are so striking in the organic world, and which Teleology has done good service in keeping before our minds, without being false to the fundamental principles of a scientific conception of the universe." Indeed: "The apparently diverging teachings of the Teleologist and of the Morphologist are reconciled by the Darwinian hypothesis" (Huxley 1893, 86). He even went as far as to say that "there is a wider Teleology, that is not touched by the doctrine of Evolution, but is actually based upon the fundamental proposition of Evolution." But be not deceived. There may still be an emaciated argument to organized complexity. However, it should not become the obsession of the biologist or the springboard of the natural theologian.

Huxley's teleology turns out to be the blind workings of law, leading in a deterministic fashion to their inevitable ends: "The whole world, living and not living, is the result of the mutual interaction, according to definite laws, of the forces possessed by the molecules of which the primitive nebulosity of the universe was composed" (Huxley 1873, 272–274). What we cannot now say is that the end to which all of this leads has any purpose. The universe is like a clock, and we have no right to say that the purpose is telling time any more than that the purpose is senseless ticking.

For Huxley, Darwin had liberated biology from the theological shackles of the past. In a real way, Huxley saw Darwin's ideas as having made the God hypothesis not simply redundant but in most respects untenable. The teleology left after the *Origin* could never (in Huxley's eyes) be strong enough to satisfy the believer. In Huxley's opinion, we have gone from support, through consistency, to opposition. Which conclusion suited him just fine!

Asa Gray

Let us cross the Atlantic and stop off in Cambridge, Massachusetts, home of Harvard University. Asa Gray (1810–1888), born in upstate New York and now professor of botany at America's premier university, had trained to be a physician. As with Huxley, this was a common first step for someone of modest means intent on a life of science. In Gray's case, he moved quickly to the study of plants, his true love and lifelong profession. Just as Huxley became Darwin's champion at home, so Gray became Darwin's champion in the New World. He was introduced to the great evolution secret before the *Origin* was published and then, when things went public, he defended Darwin's ideas from the podium. Gray's opponent was the planner of Harvard's massive Museum of Comparative Zoology, Louis Agassiz, a Swiss ichthyologist transplanted to Boston and adored by the New England intelligentsia as a scientist who properly combined the study of nature with the proclamation of the powers and achievements of the Creator.

Agassiz had been a protégé of Cuvier in the final year of the great Frenchman's life, and although the Swiss-American probably owed his largest intellectual debt to German idealism (he had been a student of the leading *Naturphilosophen,* the philosopher Schelling and the anatomist Oken), he always believed that he was carrying the banner of opposition to Lamarck and Geoffroy. As it happens, Gray the evolutionist was at least as devout a Christian as Agassiz (who had in fact moved from the Swiss Protestant faith of his childhood to the Unitarianism of prominent Bostonians). An evangelical Christian, doctrinally Presbyterian, Gray much enjoyed hobnobbing with Anglican clergyman on his frequent trips to Europe. But this was no barrier to his embrace of evolution, a function partly of his personal relationship with Darwin (whom he had met through their shared friendship with Joseph Hooker, also a botanist) and probably partly a function of his Presbyterian commitment to the study of nature and the rule of law.

The American debate over evolution centered around these two figures, but it was a public rather than a private affair—more popular science in the realm of ideas and beliefs and commitments of laypeople than serious science in the professional community. As was the case with Huxley, in Gray's professional work, in the early post-Darwinian years especially, evolution plays a minimal role and selection is vanishingly insignificant. But Gray was certainly more attuned to adaptation than was Huxley. Like Hooker, in the 1850s Gray took a keen interest in the geographical distributions of plants and the relative ratios and interchanges between continents. Matters of function were absolutely crucial here, as Gray tried to tease out the histories of plants and their ecosystems, seeing which features might be reflective of ancestry and which of adaptation to the immediate surroundings. But it was far from the case that Gray understood the power of selection when Darwin first presented it, or made selection a crucial and constant tool of botanical investigation. Anything but! Revealing is the topic of phyllotaxy, the arrangement of leaves on the stem of a plant. This can be shown to follow certain mathematical rules, much cherished by the *Naturphilosophen*, who took phyllotaxy as evidence of the underlying order and regularity of creation. In the early 1870s, Gray's fellow Bostonian, the pragmatist and evolutionist Chauncey Wright, showed that phyllotaxy can be explained through selection. But later in the decade, when discussing the topic of phyllotaxy, Gray did not mention selection, although he made full use of Wright's calculations and acknowledged his debt to Wright's work. Gray simply did not view the argument to organized complexity, and Darwin's solution of national selection, as a necessity for and obsession of the working scientist. It was a topic of broader interest, to be discussed in a collection of essays for the general public. And here, questions of teleology were not only appropriate but expected.

Gray on Function

Like Huxley, Asa Gray saw Darwin's great achievement as solving jointly the question of homology and adaptation. Quoting Darwin's claim that his theory "admirably serves for explaining the unity of composition of all organisms, the existence of representative and rudimentary organs, and the natural series which species and genera compose," Gray agreed wholeheartedly (1876, 99–100):

> Suffice it to say that these are the real strongholds of the new system on its theoretical side; that it goes far toward explaining both the physiological and the structural gradations and relations between the two kingdoms, and the arrangement of all their forms in groups subordinate to groups, all within a few great types; that it reads the riddle of abortive organs and of morphological conformity, of which no other theory has ever offered a scientific explanation, and supplies a ground for harmonizing the two fundamental ideas which naturalists and philosophers conceive to have ruled the organic world, though they could not reconcile them; namely, Adaptation to Purpose and Conditions of Existence, and the Unity of Type. To reconcile these two undeniable principles is the capital problem in the philosophy of natural history; and the hypothesis which consistently does so thereby secures a great advantage.

But this said, the differences between Darwin's two greatest supporters were major. As a believing and practicing Christian, Gray was eager to establish harmony between religious belief and science. Again and again, he stressed that there is no incompatibility between Darwin's theory and the Christian faith. "We could not affirm that the arguments for design in Nature are conclusive to all minds. But we may insist, . . . that, whatever they were good for before Darwin's book appeared, they are good for now." Gray came out four-square in favor of traditional natural theology. "To our minds the argument from design always appeared conclusive of the being and continued operation of an intelligent First Cause, the Ordainer of Nature; and we do not see that the grounds of such belief would be disturbed or shifted by the adoption of Darwin's hypothesis." Moreover, he was sensitive to the distinction between the argument to organized complexity and the argument to design, although there does seem to be some concession to the weakness of the latter. The Darwinian system "not only acknowledges purpose . . . but builds upon it; and if purpose in this sense does not of itself imply design, it is certainly compatible with it, and suggestive of it. Difficult as it may be to conceive and impossible to demonstrate design in a whole of which the series of parts appear to be contingent, the alternative may be yet more difficult and less satisfactory" (Gray 1876, 311).

In fact, however, Gray claimed that Darwin's theory in some respects strengthens the argument to design rather than weakens or refutes it. He argued that with evolution through natural selection, it was no longer necessary to search for *ad hoc* reasons why not every organic feature shows clear and unambiguous design. Under the Darwinian hypothesis, which Gray referred to as "comprehensive and far-reaching teleology," we find that "organs and even faculties, useless to the individual, find their explanation and reason of being. Either they have done service in the past, or they may do service in the future. They may have been essentially useful in one way in a past species, and, though now functionless, they may be turned to useful account in some very different way hereafter" (p. 308).

Scientifically and metaphysically also, Gray differed from Huxley. Whereas the Englishman had favored saltations—anathema to a person trying to explain adaptation through natural causes—the American thought that the very way that the world is constituted is such as to point away from abrupt change. Referring to the Leibnizian axiom, "Natura non agit saltatim" (Nature does not make jumps), Gray agreed that you cannot simply elevate it to the status of a necessary truth. Nevertheless, "naturalists of enlarged views will not fail to infer the principle from the phenomena they investigate—to perceive that the rule holds, under due qualifications and altered forms, throughout the realm of Nature; although we do not suppose that Nature in the organic world makes no distinct steps, but only short and serial steps—not infinitely fine gradations, but no long leaps, or few of them" (Gray 1876, 101).

However, finally, and here is the real rub, Gray broke with Huxley's view of a world subject only to unbroken law of a basic materialistic kind. He was not prepared to accept that Darwin had given a fully adequate answer to the argument to organized complexity, and therefore

Darwin had not offered a naturalistic alternative to the traditional deity-ending argument to design. Crucially, Gray could not see how selection alone could produce adaptation. He felt that the variations on which it worked had to do some work also. They had to be specially guided in some way (pp. 121–122):

> Wherefore, so long as gradatory, orderly, and adapted forms in Nature argue design, and at least while the physical cause of variation is utterly unknown and mysterious, we should advise Mr. Darwin to assume, in the philosophy of his hypothesis, that variation has been led along certain beneficial lines. Streams flowing over a sloping plain by gravitation (here the counterpart of natural selection) may have worn their actual channels as they flowed; yet their particular courses may have been assigned; and where we see them forming definite and useful lines of irrigation, after a manner unaccountable on the laws of gravitation and dynamics, we should believe that the distribution was designed.

Speaking in rough terms, we might say that Huxley was indifferent to adaptation. He was unconcerned—even happy—that the argument to complexity was not strong enough to require a full response to the argument to design, especially not a theological response. Happy because, as a scientist, Huxley did not really think that natural selection was all that powerful a mechanism and hence could not tackle really difficult problems. Happy because, as a secular theologian, he did not want to have to agree that some aspects of the organic world cannot be given a naturalistic explanation, especially not an explanation that makes the world something of transcendent value.

Although later in life he grew pessimistic and skeptical about any ultimate meanings, secular or otherwise, generally for Huxley evolution as such had a kind of natural value—it explained humankind and our position in nature—but he always wanted to deny values of any kind at the adaptation level. Christians saw the eye as a mark of the Christian God's beneficence, and Huxley wanted none of that. Gray agreed with Huxley that a Darwinian response could not do the trick, but—being sensitive to adaptation and thinking that the argument to complexity did need a response—Gray retained a theological version of the argument to design. He wanted natural selection supplemented by God-driven variations, thus blurring the boundary between the natural and the supernatural. He clung to values of a pre-Darwinian kind.

Darwin's Angst

What was Darwin's reaction to all of this? As you might imagine, he had little time for any of it—from Gray or anyone else, including his old friend the geologist Charles Lyell, who was likewise agonizing over the need to keep a theological perspective on origins. Darwin wrote to Lyell that "astronomers do not state that God directs the course of each comet & planet.—The view that each variation has been providentially arranged seems to me to make natural selection entirely superfluous, & indeed takes whole cases of appearance of new species out of the range of science." Referring to his favored analogy with the domestic world, Darwin continued: "He who does not suppose that each variation in the Pigeon was providentially caused, by accumulating which variations man made a Fantail, cannot, I think, logically argue that the tail of the Woodpecker was formed by variations providentially ordained."

The simple fact of the matter, for Darwin, is that you have to keep God out of it all. "It seems to me that variations in the domestic & wild conditions are due to unknown causes & are without purpose & insofar accidental; & that they become purposeful only when they are selected by man for his pleasure, or by what we call natural selection in the struggle for life & under changing conditions." Darwin was certainly not impressed by Gray's stream-of-water analogy. He thought indeed that it was like to "smash the whole affair" (Darwin 1985, 9.226–227).

Darwin was no less forthright with Gray himself, even though—writing in the early 1860s—he still saw glimmers of an overall plan behind everything (8.275):

> One more word on "designed laws" & "undesigned results". I see a bird which I want for food, take my gun & kill it, I do this *designedly*.—An innocent & good man stands under tree and is killed by flash of lightning. Do you believe (& I really shd. like to hear) that God *designedly* killed this man? Many or most persons do believe this; I can't and don't.—If you believe so, do you believe that when a swallow snaps up a gnat that

> God designed that that particular sparrow shd. snap up that particular gnat at that particular instant? I believe that the man & the gnat are in the same predicament.—If the death of neither man or gnat are designed, I see no good reason to believe that their *first* birth or production shd. be necessarily designed. Yet, as I said before, I cannot persuade myself that electricity acts, that the tree grows, that man aspires to loftiest conceptions all from blind, brute force.

Darwin did not leave things at this. He set about offering a rival metaphor to Gray's stream of favorable variation. In the *Variation of Animals and Plants under Domestication* (1868), he invited the reader to think of an architect designing and building a great structure. Of course, there is intention in the actual planning, but what about the actual stones that are used in the building? The architect does not design their shape. He just takes what he has and works from them. The situation is entirely analogous to that encountered in the organic world. "The fragments of stone, though indispensable to the architect, bear to the edifice built by him the same relation which the fluctuating variations of organic beings bear to the varied and admirable structures ultimately acquired by their modified descendants."

It is certainly true, Darwin continued, that some people (Asa Gray) seem to think that selection cannot work unless each variation is designed. But this is on a par with a "savage" (nasty blow here) insisting that the construction of the building was incomplete without an explanation—an intention-based explanation—of the shape of each stone. Of course there are laws governing the shape and form of the stones. "But in regard to the use to which the fragments may be put, their shape may be strictly said to be accidental." There is no need to bring God into all of this. "An omniscient Creator must have foreseen every consequence which results from the laws imposed by Him. But can it be reasonably maintained that the Creator intentionally ordered, if we use the words in any ordinary sense, that certain fragments of rock should assume certain shapes so that the builder might erect his edifice?" The situation is entirely similar in the organic world. "Did He ordain that the crop and tail-feathers of the pigeon should vary in order that the fancier might make his grotesque pouter and fantail breeds?" (Darwin 1868,

2.426–428). Obviously not. Why then bring him into the natural world? Asa Gray is simply wrong on this.

Gray could not leave things at this. He came back with a new metaphor of his own, intending, however, the same end result. Natural selection should not be thought of as the wind that drives a ship. It is rather the rudder that pushes things one way or the other. "Variation answers to the wind: 'Thou hearest the sound thereof, but canst not tell when it cometh and whither it goeth.' Its course is controlled by natural selection, the action of which, at any given moment, is seemingly small or insensible; but the ultimate results are great" (Gray 1876, 316–317).

Ultimately, however, there was to be no meeting point here between Gray and Darwin. Gray, unlike Darwin's other supporter, Huxley, appreciated adaptation. Gray, unlike Huxley, thought that Darwinism and Christian belief were compatible. Gray, unlike Huxley, was a gradualist. But most important, Gray, unlike Huxley, wanted to take evolution out of the realm of science, at least when it came to variation. This was the point where Darwin drew the line. It was one thing to see an overall plan to everything but quite another when that plan came filtering into the science. This was precisely the sort of teleological thinking from which Darwin was breaking.

Of course, with hindsight, we might feel that an unsatisfactory response on questions of teleology is all we would expect from people who were not prepared to make use of the gifts Darwin offered, natural selection specifically. Huxley did not want to accept the mechanism because he could not appreciate the problem. Gray did not want to accept the mechanism because he *could* appreciate the problem! Only after Darwin's ideas were employed as the essential tools of working scientists can we see how this argument would play itself out.

CHAPTER 8

THE CENTURY OF EVOLUTIONISM

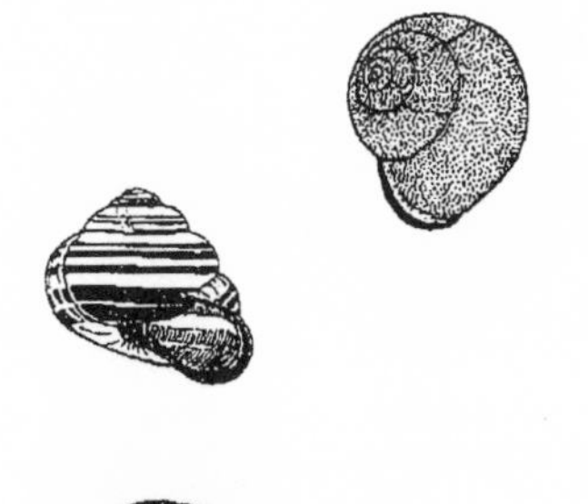

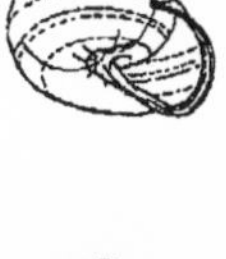

DARWIN'S DREAM of a functioning, professional, selection-based evolutionism did not materialize in the years following his death. At best, at the professional level, evolution became a second-rate, Germanized tracing of phylogenies. Its main contribution lay in other directions. Hijacked by progressivism and underpinned by social Darwinism, evolution became the secular religion of the age—a new philosophy for a new dawn.

But surely, one might argue, this sociological account is incomplete to the point of distortion: the real reason why Darwinism did not take root and flourish as a professional science grounded in natural selection was primarily one of pure science. The flaw in Darwin's theory, as presented in the *Origin,* was that he had no proper theory of heredity, of how characteristics are passed from one generation to the next. In today's terminology, we would say he had no theory of genetics. Instead of thinking, as we think now, that heredity is particulate—that the causes of heredity pass on unchanged and intact—Darwin thought, as we do not think now, that heredity blends in each generation and is therefore diluted. Rather than taking sex (which persists undiluted; offspring are either male or female) as the touchstone for what happens at the causal level, Darwin took skin color (which blends) as his touchstone. As the Scottish engineer Fleeming Jenkin and other critics of his own day were quick to point out, blending means that selection can never be effective in bringing about species change, because no matter how advantageous a new variation may be, within a generation or two it will be diluted out of existence. Until this problem was solved, there was no way for a professional, selection-based science to persist and flourish.

Well, yes and no. These two accounts—the scientific and the sociological—are not mutually exclusive; both are needed for a full understanding. Though the significance of the heredity factor cannot be denied, it cannot explain all. Before the problems of heredity were solved in a satisfactory manner, a truncated Darwinism would still have been possible. Darwin himself showed in his orchids book that it is possible to do creative, serious science using the mechanism of selection, without a full knowledge of how variations are transmitted. And one could argue that such a truncated Darwinism might even have led more rapidly to the correct principles of heredity. It is not as though these principles were out of reach. The man whom we revere today as the seminal thinker about heredity, the Moravian monk Gregor Mendel, did his path-breaking work in the 1860s; it was only because of his isolation from mainstream science that he remained unknown until 1900.

But, speculations aside, the fact is that several of Darwin's contemporaries—unlike Huxley and Gray—did take selection seriously as a research tool. To kick off our survey of a century of evolutionism, from 1859 to 1959, let us begin by looking at some of these selectionists, starting with the man who discovered mimicry and explained it by invoking Darwin's mechanism.

Henry Walter Bates

Kirby and Spence's *Introduction to Entomology* (1815–1828), a work that went through many editions, reflected the English fondness for insect collecting. Darwin himself, when at Cambridge, spent happy hours seeking, capturing, and pinning beetles, and he was one of many. Butterflies,moths, and other insects were the prey eagerly sought by hosts of ardent naturalists—it was a highly respectable leisure occupation for both sexes—and with the growth of empire, tastes expanded from the domestic and ordinary to the foreign and exotic. Many a soldier or missionary, administrator or spouse, in some distant corner of British rule spent their free time in pursuit of the brightly colored insect denizens of their locality. Moreover, rich collectors paid good money for specimens, and this economic incentive spurred Alfred Russel Wallace and his friend Henry Walter Bates to set off in the 1840s for the exotic jungles of South America.

Bates spent twelve years around the Amazon, and on his return to England put his discoveries to good theoretical use. He provided a detailed explanation of butterfly mimicry using natural selection as the investigative tool which, to this day, is still properly considered one of the triumphs of Darwinian biology. The focus of Bates's paper, published in 1862 by the Linnean Society of London (which had published the Darwin-Wallace papers in 1858), was a widespread family of butterflies found in tropical South America, the Heliconidae. Bates himself discovered 94 species, and the numbers within each species were high. "The species are exceedingly abundant in individuals wherever they occur: they show every sign of flourishing existence, although of slow flight, feeble structure, unfurnished with apparent means of defence, and living in places which are incessantly haunted by swarms of insectivorous birds" (Bates 1862, 499). Given the widespread distribution of the Heliconidae and their large numbers, naturally Bates found geographical overlap with other totally unrelated butterflies. One particular case involved members of the genus *Ithomia,* from the Heliconidae family, which overlapped with members of the genus *Leptalis,* from the Pierinae family. This set up the problem that Bates would eventually solve with the concept of mimicry. "The resemblance is so close, that it is only after long practice that the true can be distinguished from the counterfeit, when on the wing in their native forests" (p. 504).

Why does this mimicry exist? Bates asked. It is a very widespread phenomenon in the living world, certainly not unique to butterflies, as Bates himself pointed out. The answer Bates offered was couched in the language of teleology (pp. 507–508):

> It is not difficult to divine the meaning or final cause of these analogies. When we see a species of Moth which frequents flowers in the daytime wearing the appearance of a Wasp, we feel compelled to infer that the imitation is intended to protect the otherwise defenceless insect by deceiving the insectivorous animals, which persecute the Moth, but avoid the Wasp. May not the *Heliconide* dress serve the same purpose to the *Leptalis?* Is it not probable, seeing the excessive abundance of one species and the fewness of individuals of the other, that the *Heliconide* is free from the persecution to which the *Leptalis* is subjected?

Bates was speaking here of adaptation, of a kind of camouflage that is so important in the living world. "I believe, therefore, that the specific mimetic analogies exhibited in connexion with the *Heliconidae* are adaptations—phenomena of precisely the same nature as those in which insects and other beings are assimilated in superficial appearance to the vegetable or inorganic substance on which, or amongst which, they live" (p. 508).

But what is the underlying cause of this phenomenon? Since this was a scientific paper, there was no place for God, especially for a person of Bates's generation. And even in his personal belief system, he had already moved to agnosticism, which was becoming the philosophical mark of his set. The answer Bates gave to the question of causation, therefore, was not divine design but Darwin's mechanism, natural selection. The Heliconidae, the models, are apparently foul-smelling and foul-tasting, highly unpalatable to the birds of their neighborhood. They can risk drawing attention to themselves by being brightly colored, and indeed they make use of this feature to warn birds away. They have no need of fast or tricky flight; their very nature is advertisement enough. The *Leptalis,* the mimics, are comparatively rare (about 1 in a 1,000 compared with the models) and are not at all foul-tasting; but through their coloration they pretend to be. They get by through deceit. "I never saw the flocks of slow-flying *Heliconidae* in the woods persecuted by birds or Dragon-flies, to which they would have been easy prey; nor, when at rest on leaves, did they appear to be molested by Lizards or the predacious Flies of the family *Asilidae,* which were very often seen pouncing on Butterflies of other families. If they owe their flourishing existence to this cause, it would be intelligible why the *Leptalidae,* whose scanty number of individuals reveals a less protected condition, should be disguised in their dress, and thus share their immunity" (p. 510).

Natural selection was Bates's explanation for mimicry. Those *Leptalis* that were, by chance variation, similar to the Heliconidae family survived and reproduced because birds avoided eating them. Those that were not similar to Heliconidae became a meal at an early age. And in the fluctuating variations we can even see selection becoming more and more successful. There is no need to suppose any other explanation, especially not some kind of internally directed force. Bates admitted that at first glance "it seems as though the proper variation always arose in the species, and the mimicry were a predestined goal." One might even be tempted to posit an innate tendency or force pushing in the direction of some preordained goal. But Bates was not lured by that possibility: "On examination, however, these explanations are found to be untenable, and the appearances which suggest them illusory. Those who earnestly desire a rational explanation, must, I think, arrive at the conclusion that these apparently miraculous, but always beautiful and wonderful, mimetic resemblances, and therefore probably every other kind of adaptation in beings, are brought about by agencies similar to those we have here discussed" (pp. 514–515).

This really is a Darwinian approach to science in general and final cause in particular. Darwin loved the paper, as did Wallace, who then wrote a paper of his own on the topic, extending the ideas to other parts of the tropical world. But wonderful though Bates's work truly was, no one picked up on his paper to make a functioning science of it; no new discipline was founded on this groundbreaking work. Bates himself—who had spent years studying the natural world at close range, albeit lacking any formal training—was refused a job in zoology at the British Museum; they preferred a nonbiologist with literary talents. Only through the recommendation of his publisher, John Murray (to whom he was directed by Darwin), did he get a job as a paid secretary at the Royal Geographical Society, where he worked for the rest of his life. It was a rare occasion when he had time and opportunity to write on the classification of insects and evolutionary theory. Only in the 1880s was he, along with Wallace, made a fellow of the Royal Society. Huxley, born in the same year as Bates, was by then president and had been a fellow for thirty years. Brilliant observer and thinker though he may have been, Bates was out of the loop of professional science.

Ronald A. Fisher

There were other selectionists in the half century after the *Origin*, but not many. One was the Oxford entomologist Edward Poulton. Another was Raphael Weldon, one of the early biometricians (people who applied statistics to issues of heredity). But they were exceptions, and theirs was not the most influential science of the day. Far more significant were the rediscovered studies on heredity of Gregor Mendel. Unfortunately, the English champion of this work, William Bateson, saw in Mendel's new ideas an evolutionary mechanism to rival the claims of the selectionists. Entirely downplaying the significance of adaptation and concentrating on the large variations that were the focus of Mendel's studies, Bateson and his associates saw here support for a saltationist type of evolutionism, where major change comes not through selection working on minute variations over long stretches of time but all at once, through large jumps over the species barrier. Selection, according to the saltationist view, operates more to clear out the detritus of evolution than to do anything creative in its own right.

The extreme position of Bateson and his contemporaries did not last long, for soon people saw that Mendelian genetics is no genuine rival to selection. The two are complements, not competitors. Mendelian genetics showed that variations get transmitted from generation to generation in particulate fashion, without being blended or diluted away. This meant that selection could work effectively and creatively, weeding out some characteristics and favoring others. Unlike the Red Queen, running frantically on the spot merely to stay still, selection could move things forward, by working on very small variations to increase adaptation. Huge leaps of the kind Bateson promoted were not necessary and indeed were deleterious.

The crucial push toward understanding how particulate heredity works came in the second decade of the twentieth century, in the "fly room" of Thomas Hunt Morgan and his associates at Columbia University. They showed how the units of heredity (the "genes") can be located on the chromosomes within the cells, and they proved that the effects of genes can be small, sometimes near-vanishingly so. With the continuous transmission of these small variations, which these Mendelian geneticists soon realized are produced through a spontaneous change in the unit of heredity (a "mutation" of the gene), one has precisely the kind of background support that Darwin lacked in the *Origin*.

Not that all of the important work was done in the New World. Even before Morgan worked out the details in his New York laboratory, British biologists realized that a generalization of Mendel's basic principles (which became known, after its discoverers, as the Hardy-Weinberg law) gives one the background stability on which the creative forces of evolution could play. "The relative scarcity of the mutation at the start does not prevent that a number of individuals interbreeding at random, some with and others without a certain factor, will give rise to a population of impure heterozygotes and pure homozygotes in which the proportion of the three classes will be in equilibrium so soon as the square of the number of heterozygotes equals the number of pure 'dominants' multiplied by the number of pure 'recessives'" (Goodrich 1912, 69). It took another two decades, however, before population biologists would put everything together into one coherent picture.

In England, the most significant of these mathematically inclined biologists was the statistician Ronald Fisher. His *Genetical Theory of Natural Selection,* published in 1930 and a landmark in the history of evolutionary theory, integrated Darwinian selection with Mendelian heredity. During his schooling, Fisher first concentrated in mathematics and afterward took a year's graduate study on the theory of gases. This physics training always showed in his biological work, for his picture of evolution was of a large group of individuals, each carrying a distinctive batch of genes, interacting through time and giving evidence at the macro-level of holistic-type effects. Most particularly, Fisher saw that when you have genetically based variation in a population, some individuals will be performing better than others (they will be "fitter"), and so selection can push the population up to the level of the best.

Yet, in Fisher's view, the environment was constantly changing and (from the point of view of organisms) deteriorating, thus making the summit ever harder to achieve. He saw selection as a gradual, subtle process that nudges and guides species toward adaptation, rather than forcing them. "I think of the species not as dragged along laboriously by selection like a barge in treacle, but as responding extremely sensitively whenever a perceptible selective difference is established. All simple characters, like body size, must be always very near the optimum" (Bennett 1983, 88). This letter to his friend and adviser Major Leonard Darwin, a son of Charles Darwin and Fisher's sometime sponsor, provides a hint of Fisher's stance on adaptation. He was a fanatical Darwinian who saw the action of selection as supreme. Adaptation was, for him, a basic and pervasive fact of the organic world.

Fisher analyzed and discussed mimicry at length and offered his reader a strong defense of the traditional English position, namely, that selection was the all-important factor. Sentiment apart, there was a strong polemical reason why Fisher plunged into this topic, for the Mendelians—specifically Reginald Punnett, Bateson's successor as professor of genetics at Cambridge—had taken up mimicry as an example of a biological phenomenon that simply could not be given an adequate selection-based explanation. Pointing to the fact that some butterflies, fundamentally different in appearance because they mimic quite different models, come not only from the same brood but are separated genetically by just one gene—they seem to be a single mutation apart from one another—Punnett argued that mimicry could not have come about slowly and gradually through a selective process. Adaptation is not the almighty god supposed by the Darwinians: "If the new character which arises independently of natural selection is neither of service nor disservice to its possessor in the struggle for existence, there seems no reason why it should not persist in spite of natural selection" (Punnett 1915, 4).

To counter Punnett, Fisher drew on the work of the Harvard geneticist W. E. Castle, who, through selection experiments with rats, had shown that a particular physical effect, such as coat color, might appear to be the result of just one variation in a gene but in fact is controlled by not only this gene but a host of others that modify—intensifying or reducing—the original gene's effects. "The gene, then, may be taken to be uninfluenced by selection, but its external effect may be influenced, apparently to any extent, by means of the selection of modifying factors" (Fisher 1930, 166). Hence, Punnett could be absolutely right that totally different morphs are produced in the same brood, but this does not mean that only one gene is involved. There could be many pertinent interacting genes—modifying genes—on which selection can and does work its magic. The genes that produce the switch from one form to another have persisted, untouched by selection. It is the secondary genes that are shuffled and selected. "The gradual evolution of such mimetic resemblances is just what we should expect if the modifying factors, which always seem to be available in abundance, were subjected to the selection of birds or other predators" (p. 166).

Fisher's ultra-adaptationism fit well with his theological convictions. He was not just a Christian but a sincerely committed Anglican who preached in his college chapel and wrote openly on matters of theology.

He was no biblical literalist, but he certainly thought of history in biblical terms, writing that God created through evolutionary processes and that our era is right in the middle of this creative event. "To the traditionally religious man, the essential novelty introduced by the theory of the evolution of organic life, is that creation was not all finished a long time ago, but is still in progress, in the midst of its incredible duration" (Fisher 1947, 1001). Fisher thought of adaptation as essentially good and representative of God's creative intent. He wrote of the "high perfection of existing adaptation" (letter to Sewall Wright, January 19, 1931, in Bennett 1983, 279) and saw the links back from Darwin to Paley's teleological argument. For Fisher, adaptation was part of God's overall plan—it was not random. Design and Darwinism went hand in hand. To use the language of final causes, we might say that Fisher accepted the argument to organized complexity as much as did Paley and Darwin. There is adaptation everywhere, and he thought as a scientist (with Darwin) that natural selection is the way to speak to this issue. However, he did not look upon this as excluding a Christian interpretation of the world.

Ecological Genetics

In Britain, the Fisherian approach was promulgated by his great friend and often co-worker E. B. (Henry) Ford. Ford, who spent his whole professional life as a student, lecturer, and professor at Oxford University, founded a school of "ecological genetics." Although he was himself an excellent naturalist and skilled observer, Ford's real strength proved to be his ability to attract a group of younger evolutionists whose work would push forward a Darwinian program, reinvigorated by Mendelian genetics, into related fields. Their focus was the application of Mendelian genetics to groups of individuals—so-called population genetics.

Ford and his followers—notably A. J. Cain, H. B. D. Kettlewell, and P. M. Sheppard—homed right in on problems of adaptation and of the putative selective causes behind it. Bates would have felt very comfortable in their group; Ford himself was a butterfly man. But probably the most brilliant work was done by Cain and Sheppard on the different forms (polymorphisms) in British snails, especially *Cepaea nemoralis.* Whenever one looks at populations of these snails, at least two strikingly different color patterns and markings become apparent. These turn out not to be randomly distributed. Specifically, "in woods where the ground is brown with decaying leaves, and fairly uniform in appearance, the unbanded brown or the unbanded or 1-banded pink shells are particularly common, whereas in hedgerows or rough green herbage the yellow 5-banded form tends to be the commoner" (Sheppard 1958, 86). Why was this? The answer was bird predation and the selective advantage of some colors and patterns over others. Because thrushes smash open the snail shells on stones, one can count the relative proportions of victims. The finding was that "in certain rough habitats the banded forms are at an advantage to the relatively unbanded ones, whereas when the background is uniform the effectively unbanded ones are at an advantage. It has also been found that the yellow forms (which with the animal inside often have a greenish tinge) are less predated on green backgrounds, and that on brown ones the pinks and browns are advantageous" (pp. 86–88).

Ford and his school worked on the argument to organized complexity—nothing more and nothing less. This was the British tradition, going back before Darwin to Kirby, Spence, and earlier naturalists. Mimicry, and adaptive coloration more generally, was the paradigm of a natural phenomenon in need of—and susceptible to—explanation. And even when the explanation was evolutionary, it was couched in highly anthropomorphic terms (Sheppard 1958, 152):

> It is obviously advantageous for an organism which is relatively inedible or dangerous to advertise this fact as widely as possible, so that it is not repeatedly attacked and thereby damaged. Consequently colours and patterns are evolved which are conspicuous and easily remembered. It is not surprising, therefore, that the patterns tend to be simple and the colours frequently include red, yellow, black or white which, it will be noticed are also used for road signs as they stand out well against natural backgrounds.

The language and thinking were that of design—of intentionality, of functioning. Animals are faced with a problem, and they come up with a solution, namely, being unpleasant. But they need to sell the sizzle as well as the steak—it is no advantage to the prey if a predator discovers this unpleasantness after he has taken a big bite out of his victim's backside. The prey must "advertise" his unpalatability to the prospective diner. The way to do this is with stark colors, which convey information quickly and unambiguously. Brightly colored road signs are not media for transmitting philosophical niceties, and neither are the markings of poisonous insects.

All very English. But what about those not squarely in that tradition? What about those in the land whose world dominance in the twentieth century would rival that of the British Empire in the nineteenth? Along with America's material and physical success, its science came to maturity in the early decades of the twentieth century and began to outstrip all others, in evolutionary biology as well as the physical and social sciences.

Adaptive Landscapes

Pragmatism was the most famous American philosophical school of the late nineteenth century, and some of its members, such as Chauncey Wright, became Darwinians. Generally, however, in the United States as in England, while people became willing evolutionists, they did not become selectionists. The adaptationist approach taken by the *Origin* was a nonstarter. Only with the growing acceptance of Mendelian heredity and population genetics did biologists begin to understand the implications of adaptation for evolution.

America's brilliant contribution to the cause came from Sewall Wright, who was born in the Midwest and trained at a little Universalist college where his father was a professor. He went on to Harvard for graduate work, studying under Castle. Wright's first job was with the United States Department of Agriculture, where he worked for ten years on problems of animal breeding, particularly on how Shorthorn cattle could be maintained and improved through inbreeding. Always interested in evolution, Wright put together his own version of evolutionary theory, which he published at the beginning of the 1930s, five or six years after he had moved to an academic post at the University of Chicago.

Formally, the Englishman Fisher and the American Wright agreed on the mathematics, even though they used very different mathematical techniques. Fisher had classical training and it showed; Wright was a pragmatist and it showed; but the calculations came out the same. Nevertheless, the two men's overall conceptions of evolutionary change could not have been more different. Fisher always saw natural selection as the determining factor in evolutionary change. For him, mutations came into large populations and either failed to make the grade or produced variations superior to those already existing and so rapidly became the norm throughout the group. For Wright, with cattle-inbreeding successes vividly before him, the key to change was the fragmentation of populations into small groups (presumably thanks to external causes). It is in these small groups, with their restricted gene pool, that real innovative change takes place.

The crucial part of Wright's theory was that such change is for the most part a result of random factors—the vagaries of breeding. Although one form A may be biologically superior (fitter) to another form B, random mating in a small group, which will invariably have some coincidences and distortions, might mean that B will prevail nevertheless—its genes will "drift" to fixity (to being the only representatives in the population). Then, as population barriers break down, these new B-type features, which themselves may not be directly adaptive, will join the general population as subpopulations within the main group. In the full-population situation, the new features may have advantages and thus get selected and fixed through the whole population. Wright's point is that they have got to come into existence in the first place, and it is isolation and drift that bring this about. Selection, Fisher-style, in a large interbreeding group could never do the job. One would never get the kinds of gene complexes or combinations that make for new features, held together for enough time to achieve the kind of integration and stability needed for them to succeed in such large groups.

Wright called his theory the shifting balance theory of evolution, and in support of his ideas, Wright introduced a powerful metaphor: he suggested that we consider organisms to be on an "adaptive landscape," with successes at the tops of peaks and failures down in the valleys; evolution is a matter of moving from one peak to another (1932, 68):

> Let us consider the case of a large species which subdivided into many small local races, each breeding largely within itself but occasionally crossbreeding. The field of gene combinations occupied by each of these local races shifts continually in a nonadaptive fashion . . . With

> many local races, each spreading over a considerable field and moving relatively rapidly in the more general field about the controlling peak, the chances are good that one at least will come under the influence of another peak. If a higher peak, this race will expand in numbers and by crossbreeding with the others will pull the whole species toward the new position. The average adaptiveness of the species thus advances under intergroup selection, an enormously more effective process than intragroup selection. The conclusion is that subdivision of a species into local races provides the most effective mechanism for trial and error in the field of gene combinations. It need scarcely be pointed out that with such a mechanism complete isolation of a portion of a species should result relatively rapidly in specific differentiation, and one that is not necessarily adaptive.

Wright did not want to deny that natural selection operates and that adaptive features significant to organisms arise, but clearly he thought that many features become established in a population without any connection to survival and reproduction. "That evolution involves non-adaptive differentiation to a large extent at the subspecies and even the species level is indicated by the kinds of differences by which such groups are actually distinguished by systematists." Apparently only when you arive at the subfamily or family level do you start to get adaptive difference. "The principal evolutionary mechanism in the origin of species must thus be an essentially nonadaptive one" (Wright 1932, 168–169).

Theodosius Dobzhansky

Wright's shifting balance theory, with its adaptive landscape metaphor at its center, was the inspiration for the work of the fruit fly geneticist Theodosius Dobzhansky. He was the most important American evolutionist of the twentieth century, for all that he was born and educated in Russia and in many respects remained deeply indebted to the land of his birth to the day of his death. His *Genetics and the Origin of Species*, published in 1937, was the foundation from which the next generation of evolutionists was able to move forward and develop a full-blown discipline of evolutionary studies.

At the beginning of his book, Dobzhansky made a fundamental division that would structure his discussion. "Description of any movement may be logically and conveniently divided in two parts; statics, which treats of the forces producing a motion and the equilibrium of these forces, and dynamics, which deals with the motion itself and the action of forces producing it" (Dobzhansky 1937, 12). Evolutionary statics, as Dobzhansky understood it, consisted of basically the principles of genetics—the gene and its nature, the chromosome and its place and function within the cell, and such matters. Dobzhansky had moved from Russia in the late 1920s to work in the laboratory of Thomas Hunt Morgan, and the first part of Dobzhansky's book therefore reflected his cutting-edge knowledge of then-modern thinking on the nature of heredity. Particularly, Dobzhansky was able to draw on work that he and A. H. Sturtevant had done on chromosomal variations in fruit flies *(Drosophila)*.

After presenting this background, Dobzhansky moved to what he termed evolutionary dynamics. Here, he drew heavily on his Russian training, particularly his knowledge of variation in natural populations (Dobzhansky had specialized in lady bugs), but the underlying structure, the theoretical skeleton on which he was putting empirical flesh, was Sewall Wright's shifting balance theory of evolution. Following Wright, Dobzhansky believed that any really significant evolution must involve the fragmentation of populations, change within subgroups, and then new characteristics reaching out and moving through the population as a whole.

Wright's theory of genetic drift was not very Darwinian, and natural selection did not play an overwhelming role. The same was true for Dobzhansky. Like many in the 1930s, he was far from convinced that adaptation was as powerful a force in the biological world as people like Fisher assumed. This skepticism fit in with what Dobzhansky had learned from Wright. "The differentiation of a species into local or other races may take place without the action of natural selection." All you need is the breakup of a group and time to form new groups. "This statement is not to be construed to imply a denial of the importance of selection. It means only that racial differentiation need not necessarily or in every case be due to the effects of selection" (p. 134). Dobzhansky truly did not want to deny selection, or that adaptation is a feature of the organic world. It is just that he did not think it overwhelming or (as he stressed) something with some special significance. "Whether the theory of natural selection explains not only adaptation but evolution as well is quite another matter. The answer here would depend in part on the conclusion we may arrive at on the problem of the relation between the two phenomena. No agreement on this issue has been reached as yet" (Dobzhansky 1937, 150–151).

The fact is that *Genetics and the Origin of Species* is not a very Darwinian book. This is not a criticism: it is not anti-Darwinian by any means, but it is not focused on the problems that obsessed Darwin and Darwinians like Fisher. For Darwin, the big biological problem is adaptation. For Dobzhansky, the big biological problem is variation. Adaptation is an issue, it is important, but it is not the key factor in evolution. Dobzhansky discussed mimicry, but his treatment was cool and noncommittal. Indeed, for anyone schooled in the tradition of Bates and Fisher, the indifference is positively jaw-dropping. "Taken as a whole, an unprejudiced observer must, I think, conclude that an experimental foundation for the theory of protective resemblance is practically non-existent" (Dobzhansky 1937, 164). Dobzhansky was not being careless or offhand; it was just that this was not his sort of problem. Differences between organisms, what keeps them from interbreeding, how variations spread—these are the things that engaged Dobzhansky as an evolutionist.

One might add, parenthetically, that although Dobzhansky was no less committed a Christian than Fisher, his faith came from the Orthodox tradition, and British natural theology was never a motivating factor. The quest for signs of the Creator's designing intelligence left him cold. The theological implications of evolution for the status of human beings was the real issue for Dobzhansky.

Selection Intensifies

Shortly after *Genetics and the Origin of Species* first appeared, Dobzhansky's somewhat distant attitude toward selection and adaptation began to change, quite significantly. Helped by Wright with the theoretical aspects of his experiments, Dobzhansky conducted a long series of studies of variation and change in evolution, using the fruit fly as his model organism in nature as well as in the laboratory. He soon discovered that variations in chromosome structures within and between populations, which he and his fellows had taken to be paradigmatic examples of genetic drift, simply had to be reinterpreted in terms of natural selection. In particular, seasonal variations (moving from one chromosome form to another and then back again over the course of seasons), when seen in isolated populations, cannot be pure chance but rather must be ascribed to adaptive advantage: one chromosome form is favored at one season, another form at another season. Soon this was confirmed by simulated changes occurring in experimental populations kept in the laboratory. Natural selection was a lot more pervasive and powerful than Dobzhansky had realized.

This new respect for selection was not unique to Dobzhansky. Like Ford in England, Dobzhansky was brilliant at building a school around himself—a group that would take his research program and push it beyond the dreams of its founder. Students flocked to Dobzhansky, and many of these people are still influential today. Dobzhansky went out of his way to bring in other mature evolutionary biologists from other areas—botany, systematics, paleontology, and the like—and to encourage them to work with him in founding a field or discipline that could be lodged in the universities, win grants and attract students, publish journals and books. Prominent supporters of this new "synthetic theory of evolution" (the English version was more commonly known as "neo-Darwinism") were Ernst Mayr, an ornithologist and systematist (author of *Systematics and the Origin of Species,* 1942); George Gaylord Simpson, a paleontologist (author of *Tempo and Mode in Evolution,* 1944); and G. Ledyard Stebbins, a botanist (author of *Variation and Evolution in Plants,* 1950). For all of them, Wright's shifting balance model with its adaptive landscape metaphor was the conceptual background, and all of them followed Dobzhansky in becoming more adaptationist in the 1940s and 1950s.

Entirely typical was Simpson's discussion of the evolution of the horse (his own specialty was mammalian evolution). He considered two adaptive peaks, one for grazing and one for browsing. He showed how the horse started as a browser and then moved in part across to grazing, after which the browser went extinct. Initially, for unrelated adaptive reasons, the horse family members started to increase in size. This set up a secondary reaction, as "the browsing peak moved toward the grazing peak, because some of the secondary adaptations to large size . . . were incidentally in the direction of grazing adaptation." Being neither fully browser nor fully grazer was an unstable situation. Some animals moved rapidly toward grazing. "This slope is steeper than those of the browsing peak, and the grazing peak is higher (involved greater and more specific, less easily or reversible or branching specialization to a particular mode of life)" (Simpson 1944, 93). Other animals pulled back to browsing and then, for other reasons, died off completely. An adaptively fueled evolutionary shift had occurred.

As World War II receded, scientific intercourse between the group around Ford and the group around Dobzhansky increased; for example, Sheppard became a post-doctoral student of Dobzhansky and Cain of Mayr. The butterfly wing patterns and shell stripes and colors they had discovered started to have an influence on the western side of the Atlantic. Adaptation was central, and selection was the answer.

Hence, somewhat artificially taking the anniversary of the *Origin of Species* in 1959 as our benchmark, we can fairly say that, over the course of a century, a functioning, professional evolutionism—the kind that Darwin had wanted, with natural selection as the causal mechanism—had itself evolved. The problem of organized complexity was recognized as the key question that needed to be tackled, and the Darwinian approach was understood to hold the key. At this level, values were relativized to the situation, and no presumption was made about something being better on an absolute—human or divine—scale. It had taken a long time, but at last the Darwinian Revolution seemed to be complete.

CHAPTER 9

ADAPTATION IN ACTION

LOOKING BACK with hindsight to 1959, we know now that fundamental new ideas were about to be proposed in biology and major empirical research programs were about to be launched. New directions would be explored in the human realm, and with these advances and explorations would come dissent and criticism, as old antagonisms to selection and adaptation resurfaced with a vengeance. Philosophers would also start to take a serious and sustained interest in evolutionary biology. For most of the twentieth century, the physical sciences had attracted the bulk of philosophical attention. Biology was ignored, especially by philosophers in the analytic tradition, such as Bertrand Russell and Rudolf Carnap. Now, slowly at first but then with gathering speed, philosophically minded biologists and biologically interested philosophers were looking at evolutionary theory with a critical but constructive eye—excavating its foundations, examining bridges to other fields, and exploring its wider ramifications.

At the third point of the triangle with science and philosophy, religious leaders showed new interest in Darwinism and its implications. Partly, this was in reaction to the attacks on evolution by American biblical literalists and the perceived need to develop a reasoned response. Partly (as with philosophy), this was the realization that evolutionary biology itself was in ferment, making terrific advances while stirring up new and old controversies. Partly, this was a function of theology itself, and its efforts to offer an adequate account of the relationship between faith and our understanding of nature.

It is to these various developments that I turn now, believing strongly (as a fervent evolutionist) that the only way to understand them properly and fully is in the light of their own history. A good place to start our survey is at the molecular end of the spectrum. As with the coming of Mendelism, initially evolutionists feared that Watson and Crick's discovery of the DNA double helix in 1953 would be all-conquering. Who would bother with whole-organism studies when the micro-world of the gene seemed to hold all of the interesting questions and solutions for heredity, variation, and development? These worries were short-lived, for by the mid-1960s evolutionary biologists were beginning to realize that a molecular approach could be at least as much of a friend as an enemy to their project. If one stopped thinking of molecular biology as a substitute for evolutionary thought and considered it rather as an aid and an ally, opening up powerful new techniques, then exciting new vistas, both theoretical and experimental, could come into view.

Particularly exhilarating and immediate was the work done by a former student of Dobzhansky, Richard Lewontin, who found ways to use molecular biology to ask and answer hitherto impregnable questions about the amount of naturally occurring variation in populations of organisms. Exploiting the fact that cellular products have different electrostatic properties, through the technique of gel electrophoresis Lewontin and his fellow workers soon found massive amounts of variation in virtually every group studied, including (especially including) our own species, *Homo sapiens.* This was just the beginning, and as evolutionists rushed to explore genetically linked variations at the molecular level, they were able to ask penetrating questions about causes.

Fruit Flies with a Taste for Alcohol

Fruit flies have continued to be a favorite subject organism of molecular evolutionists, and one particularly well-studied topic has been that of the gene that produces enzymes enabling the insect to digest and utilize alcohol. A huge amount is now known about this alcohol dehydrogenase *(Adh)* gene, starting with its position in the genome—it is found on the left arm of chromosome 2 in *Drosophila*—and its exact chemical composition. Its molecular variants (alleles) have been described, including two particularly widespread variants, so-called fast and slow *Adh* (*Adh-F* and *Adh-S*). These variants turned out to play a crucial role in *Drosophila*'s adaptation.

Regular fruit flies, without the alcohol dehydrogenase gene, are killed by alcohol. So a variant gene that gave a fly the ability to resist—to utilize even—alcohol has intriguing adaptive possibilities, and it was not long before evolutionists turned their attention to this gene. And sure enough, it became readily apparent that the *Adh-F* gene was functioning to provide populations with otherwise unobtainable resources. The most obvious places to test out suspicions about the effectiveness of the gene are areas where alcohol is present in quantity and is prone to making its presence felt in the air, namely wineries. Sure enough, researchers found that if they went down to a wine cellar and checked around the fermentation vats, they could readily bring in a large haul of fruit flies. But not flies of any kind or species—rather, exclusively those carrying the *Adh* gene. Two very closely related species of *Drosophila* are most revealing. *D. melanogaster,* easily able to synthesize alcohol, is found within wineries. *D. simulans,* lacking the *Adh* gene, is widespread outside wineries but virtually absent within.

Can one move on from these straight observations in nature to see if experimentation can tease out and confirm some of the pertinent issues? Given that some species of *Drosophila* are better able to digest alcohol than are others, can one show that on foodstuffs that have alcohol content, one does indeed get—either as a matter of choice or of survival—those with the digestion ability more frequently than those without? Australian experimenters laid out traps containing different foodstuffs, some with vegetables unlikely to ferment, some with intermediate fruits and vegetables, and some with fruits like grapes and apples that are given to fermentation and known to produce alcohol. The findings of H. L. Gibson and his colleagues suggested strongly that fruit flies lacking the *Adh* gene were not generally to be found in great proportion in alcohol-producing contexts, whereas those that did have the gene were. Five species known to live together were studied, and different preferences for vegetables and fruits were clearly shown by the different species. Since these preferences showed a strong positive connection between fermentation potential and possession of the *Adh* gene, John G. Oakeshott and his colleagues felt justified in concluding that, more than just correlation, there is a causal link here.

What about experimentation that tries to simulate the effects of selection? John McDonald and associates attempted precisely this. They started with two identical populations of *Drosophila melanogaster.* One population was subjected, over twenty-eight generations, to intense (artificial) selection for alcohol tolerance. Those that survived alcohol environments were used to breed the next generation. The other population was not selected and was used as a control. Would selection have an effect, and would it produce an adaptive ability to survive in alcohol-containing environments and to metabolize alcohol? The results were dramatic, especially as the alcohol levels were increased. At relatively low alcohol concentrations (about 8 percent), almost all of the flies, selected and nonselected, survived. But after three days at high concentrations (about 18 percent), virtually none of the nonselected flies were still alive, but nearly three-fourths of the selected flies were still in business.

Moreover, McDonald and colleagues were able to show that these differences were directly correlated to (and only to) the enzyme produced by the *Adh* gene. In other words, they showed experimentally that the hypothesis that those fruit flies able to utilize environments like wineries with high alcohol content could quite well have been selected to do so and that the *Adh* gene could well have been a crucial causal factor.

Other questions have been raised and answered, for example about the implausibility of suggesting that the variation between *Drosophila* species with respect to alcohol tolerance is more a matter of history, what one might call "phylogenetic inertia," meaning that something possibly adaptive in the past now is simply carried along in the present. Herve Mercot and associates ruled out that hypothesis when they showed that relatedness does not correlate with adaptive ability to utilize alcohol. When flies have the chance to colonize wineries and breweries, they do and they do so quickly. The species that breed and live in human-made fermenting resources belong to three different subgenera. They did not have a shared immediate evolutionary past.

Finally, one might expect that the fast allele of the *Adh* gene would spread throughout the population, but this is not necessarily the case. Not every environment is alcohol-rich, and if there are costs to the fast gene and benefits to the slow gene, then other patterns might result. Would these patterns necessarily reveal adaptation? They would if there were systematic geographical change, "clines," showing that random factors simply could not have produced the end result. A good example is the shift from dark to light skin color in humans as populations live farther from the equator. Whatever the causes, it cannot possibly be random genetic drift. To produce such steep clines, selection has to be operating and adaptation has to be involved. Likewise in the *Drosophila* case, one finds that there are systematic changes in the frequencies of the fast and slow alleles. As one gets farther from the equator, the slow allele declines and the fast allele increases—this holds in both the northern and the southern hemispheres. Precisely why this pattern occurs is a matter of some debate. It may be linked with the fact that the slow-allele homozygote is more stable at higher temperatures, but other factors are almost certainly involved. Rainfall and humidity have been fingered as possible causes, not surprisingly since *Drosophila* species are known to be highly sensitive to these ecological factors.

The *Adh* gene in fruit flies has an adaptive function, namely to enable its possessors to utilize otherwise dangerous or lethal resources. Alcohol tolerance and digestion have an adaptive value, giving selection a substrate on which to operate. This is shown by studies in the wild as well as by experiments, both seminatural and more contrived. As the clines of *Adh* variants show, however, there is more to the *Adh* story than simply alcohol tolerance, and there are ongoing questions waiting to be answered.

Guppies and Their Predators

Ecological geneticists like Cain and Sheppard, whom we have already met, work with relatively fast-breeding organisms in an attempt to capture the forces of selection at work in nature and in the laboratory. From a host of such studies, I have chosen one that has now been ongoing for over two decades, a project focusing on little fish to be found in the rivers on the island of Trinidad. David Reznick and his colleagues and students have been much interested in the evolution of guppies, concerned particularly with the adaptive responses that these fish show with respect to predation. The background to Reznick's work is a basic theory which holds that if predation falls more heavily on the mature organism than on the young, an adaptive shift toward earlier maturity and an increase in reproductive effort would occur. The reasons for this are fairly obvious, namely, that if the adult is subject to grave danger, then the sooner and more prolifically that adult is able to reproduce, the better chance its genes will have to be passed on to the next generation. This holds even if the individual offspring produced are more disadvantaged and therefore less likely to survive than if they were born later to more mature parents. But does nature conform to the theory?

The Trinidad guppies are an excellent subject for study of this question. They occur in rivers falling on both sides of a mountain range, some of which flow north and others south. The rivers on both sides of the range are often broken up into isolated stretches because of waterfalls and other hazards. These stretches tend to be relatively similar in temperature and food supplies, but within them the guppies are at risk from different predators with different practices and preferences. The further upstream the stretch, the less voracious the predator and the more likely it is to feed on the smaller juveniles. Downstream, the predators come in from the sea and tend to be large; consequently, they can feed on full-grown adults. On the north side of the mountain range, the most significant of the downstream predators are goby and mullet; further upstream, the smaller killfish and various kinds of prawns tend to go more for the weaker juveniles and tend to be more omnivorous in their diets. The southern rivers follow a similar pattern, although with a different set of predators. Cichlids are among the really heavy predators, as well as members of the family Characidae.

Reznick started by simply looking at the fish in the various isolated stretches and seeing if their rates of maturity did indeed correspond as expected to the respective predator pressures. Reznick also did laboratory breeding tests—rearing captured guppies in standard conditions—to make sure that he was dealing with real genetic differences and not just environmentally caused variation. He also compared the findings of the northern rivers with those of the southern rivers, to cross-check that he was dealing with phenomena that were reproducible and showed, systematically, similar effects to similar causes. And what Reznick found was that theory and evidence fit almost exactly.

The greater the predation on adults, the more there was a tendency to mature early and quickly and to give birth sooner rather than later. Better to be in the game at all, picking up a base or two at a time, than to conserve one's energy for a possible home run later. Less metaphorically, even producing less-than-perfect offspring earlier was a better option than waiting for perfection and running the risk of not producing offspring at all. This was not a chance or aberrant finding. Breeding tests showed unequivocally that the differences between fish was genetic and persistent. Adaptation was the governing principle, and just doing better than the alternatives was good enough.

Reznick warned, however, that the study could not stop there. He had certainly established that there is a pattern of the kind that he sought, but was there truly a causal connection? To pursue this question, he employed a "mark-recapture" technique: he caught all of the guppies in a particular location, marked them in some definite way so that they would not be confused with others, and then released them back into the stretch. Later he caught them all again—at least all of those remaining—and thus could do comparative checks on the rates of predation. He could actually see if indeed the predicted prey were being eaten by the predators. Reznick somewhat triumphantly concluded that his hypothesis had been borne out and that a causal connection seemed plausible. In particular, where predators go for older more mature guppies, the chances of such a guppy surviving to be recaptured is significantly less than in a stretch where the predators are more interested in younger, less mature fish. Just what one expects.

But Reznick was still not satisfied. Another way of approaching the problem was by manipulating the mortality rates in nature, to see if the guppies responded in expected ways to changed selective forces. He moved guppies from high- to low-predation localities, and in another case introduced more voracious predators (of adults) into low-predation localities and then traced the evolutions of the guppies over several years and many (twenty-five or more) generations. Again, the evidence tracked the theory, with the more-predated guppies moving to earlier maturity and greater fecundity and the less-predated guppies moving in the opposite direction. Life history is under the control of selection. Adaptation is the key principle at issue.

Reznick points out that probably other factors need to be uncovered and discussed—questions having to do with population densities of the guppies, for instance. But the overall conclusion is clear and firm. When all of the studies are taken together, "they make a strong case for guppy life histories being an evolved response to prevailing mortality rates and for predators being a major cause of these differences in mortality rates" (Resnick and Travis 1996, 272).

The Sex Life of the Dunnock

Darwin located behavior, social behavior in particular, firmly within the scope of evolution through natural selection. Yet for a century the study of the evolution of social behavior lagged behind other fields like biogeography and paleontology. In part, this was because behavior is inherently so very difficult to study. You cannot just kill your organism and study it at leisure in the laboratory. In part, this was because the rise of the social sciences cast a chill on the prospect of evolutionists doing their own studies of social behavior, in animals or humans; this issue was now the property of others. And in part this was because no one could get a proper theoretical grasp on the problems to be investigated. Social behavior necessarily demands that one think in terms of groups, but, as Darwin and his more perceptive followers could see, group selection is no true answer to what is happening. And individual selection seems insufficiently powerful—except in groups that are so well integrated that they function as one organism. This was perhaps true of some hymenoptera, as Darwin realized, but obviously was not generally the case.

Things were to change dramatically in the 1960s. In opposition to the excesses of those (especially Vero Wynne-Edwards) who confidently assumed that adaptations can work for the group against the interests of the individual, a number of more thoughtful Darwinians—notably the American ichthyologist George Williams (1966)—drove a stake right through the heart of traditional group selection. At the same time, a whole new set of theoretical models was devised for dealing with social behavior. Although these models treated organisms that lived in groups, they stuck resolutely to the conviction that selective advantage must accrue to the individual. Adaptations must increase the genetic representation of their holders, and only tangentially if at all the representations of others. To use the felicitous metaphor of Richard Dawkins (1976), the models took a "selfish gene" attitude to selection and adaptation.

The most successful models were devised by the Englishman William Hamilton, who as a graduate student saw how to use an individual-selection perspective to solve the hymenoptera sterility issue—to explain, that is, the adaptive significance of all those sterile females. Much influenced by the ideas and approach of Fisher, Hamilton pointed out that, at the bottom line, selection really means increasing the relative number of copies of particular genes in the next generation. Behavior, morphology, and everything else is just a means to this end. Hence, a counterintuitive behavior, like an organism helping another in the struggle for existence, is perfectly possible so long as the genes that cause this "altruism" thereby increase their numbers in future groups. But how did this apply to sterile worker bees?

The real crux of the matter, as Hamilton pointed out, is that hymenopteran sexuality is peculiar—females have both mothers and fathers (hence one half set of chromosomes from each parent), whereas males have only mothers (hence only one half set of chromosomes in total). During her nuptial flight, the queen gets impregnated once and for all, holding the sperm in a separate compartment for the rest of her life. When she ovulates, if the egg is fertilized by one of the sperm, the resulting offspring is female; if the egg is not fertilized, then the resulting offspring is male. One consequence of this is that females share more genes in common with their sisters than they would with their daughters, if they had daughters. This follows because (by straightforward Mendelian principles) sisters share one half of the genes on their respective maternal chromosomes (half of 50 percent), but since their fathers have only one chromosome to give, they share all of the genes on their respective paternal chromosomes (all of 50 percent). This makes sisters 75 percent related, whereas mothers and daughters are only 50 percent related, just as in humans. Hence, genetically speaking, it pays a female to sacrifice her own reproductive activity to the cause of raising fertile sisters. A sterile worker puts more copies of her own genes into the next generation through fertile sisters than she would through fertile daughters. Far from being a handicap, sterility can be a very effective adaptation for its possessor.

Hamilton's mechanism—since labeled kin selection—was one of a number of new suggestions for producing social behavior through natural selection. Particularly powerful theoretically were new models using game theory, that branch of mathematics developed during World War II which investigated the alternative strategies participants (combatants) might take in order to maximize benefits to themselves, given that others were pursuing similar ends. And along with the theory came the naturalists and experimenters, trying at last to see if animal behavior could be integrated fully into the Darwinian picture. As a prize example of work in this evolutionary sub-branch, I pick the work of Nicholas Davies at Cambridge, who performed detailed studies on the tiny birds called dunnocks, perhaps better known as the hedge sparrow (see page 171).

On the surface, these birds appear to be quiet, respectable little things, going about their conventional lives with diligence and probity—working hard together as pairs to raise their demanding offspring. But closer inspection shows a riot of sexual behavior, equaling anything to be found in the pages of *Playboy* or the mind of an adolescent boy. There are indeed monogamous dunnocks. There are also polygynous dunnocks (one male, several females). There are polyandrous dunnocks (one female, several males). And there are even those referred to primly as polyandrynous, meaning that they are into group sex, with several males mated to several females.

Various factors account for this diversity of sexual practice, although the major question about why dunnocks specifically should have gone this way is still unanswered. From a Darwinian perspective, what one expects to find is that the birds' behavior is adaptive, that (in game-theory terms) they are maximizing their reproductive opportunities and payoffs by their behavior. This would mean—given, as is usual with birds, that both sexes are involved in parental care—that a male would put in effort with the young only inasmuch as he was actually certain that he was in fact a biological parent of these young, and that the proportionate amount of time and effort would be given in light of his expectations of being a biological parent. Conversely, females would allow reproductive access by the males to suit their own ends. A female in a polyandrous relationship would find it in her interest to let both males have reproductive access, thereby ensuring that both males then would work to feed and raise her young. In the Darwinian world, nobody willingly does something for nothing.

Davies and his coworkers found that the birds they studied in the

Cambridge University Botanical Gardens seemed to have very definite patterns of reproductive and offspring-raising behavior. A more dominant (alpha) male would seem to give more to the offspring than a less dominant (beta) male, and females would seem to give at least some reproductive access to all of the males with whom she was involved. Birds copulate very rapidly, but—almost as if taking pity on nosy evolutionists—dunnocks have a rather complex precopulation pattern whereby the male gets the female to eject sperm from her previous mating, thereby raising his chance of actually fertilizing the female and her chances of getting him to remain around to help with offspring.

Direct observation, however, is but a rough guide. Are the offspring really those of the caring males, and in what proportion? To answer this definitively, one needs more reliable information, which, before the molecular revolution, would have been impossible to obtain. DNA fingerprinting opened up a whole new dimension to testing hypotheses about parenthood, and the predicted connections were shown to hold almost exactly. "DNA fingerprinting showed that mating access was a good predictor of paternity. Over 99% of the young were fathered by resident males. Where both alpha and beta males mated with a female, paternity of the brood was often shared, with alpha males fathering on average 55% of the young and beta males 45%. A beta male's paternity share increased with his share of mating access." The females were just as revealing in their negative way. "In all cases maternity was assigned to the resident female on the territory" (Davies 1992, 130).

Adaptation rules, but why? Obviously, no one claims that the dunnocks are masters (or mistresses) of probability theory; but through some fairly simple experiments using fake eggs and removing males for limited periods, Davies was able to find that the actual laying of eggs seems to trigger the behavior in males. If a male was removed from a relationship before the eggs appeared, then even though he had had sexual congress with the female, he would not feed the young if later he was reinstated. On the other hand, the sight of an egg—even a model egg—before removal was generally enough to spur the male to feeding action when later he was reinstated.

Everything does not always work absolutely perfectly, of course. For a start, one is working with rough-and-ready rules that might break down if environmental factors overwhelm them. In this case, for instance, if the birds are removed too early before real egg laying, then no number of fake eggs will stimulate them into helping later. Then again, remember that we have different birds competing against one another for limited resources. It may be that one is triumphing over others. This, after all, is what selection is all about. Although beta males apportion their behavior according to their apparent rewards, in fact some are doing better genetically than they should. This could be because of their actual behavior, or because of some inadequacy (like infertility) in their alpha competitors, or just plain good luck. All of which goes to show that although adaptation rules, it does not always rule completely triumphant for all. In Davies's own words (p. 9), there are important issues about how "well designed" we might expect to find organisms to truly be. Darwinian selection works through competition and success, not perfection.

Keeping Cool

Nothing has captured the popular imagination quite like those lumbering, reptilian brutes of the past, the dinosaurs. From the Great Exhibition in London in 1851, when crowds came to gawk at fabulous (and, as we now think, somewhat fictional) reconstructions of these prehistoric giants, to the present, when the wonders of modern technology re-create them for Hollywood blockbusters, the dinos continue to thrill and terrify, to mystify and marvel. You cannot visit one of the world's great natural history museums—the British Museum in South Kensington, London, or the American Museum off Central Park, New York—without standing in awe before their enormous skeletons. Giving the lie to the popular myth that the dinosaurs were ungainly monsters, ill-adapted for life, we know now that they flourished and diversified, covering the earth. Our mammalian ancestors, tiny rodent-like creatures, survived only by making themselves inconspicuous and by sensibly being active mainly at night when their rulers slept. Indeed, the dinosaurs lasted for nearly two hundred million years, their span ending only some sixty-five million years ago when an asteroid or comet smashed into the earth and finally destroyed them. And even that is not really true, because it seems highly probable that their descendants continue to thrive and reproduce—only now we call them the birds.

Conspicuous even among the dinosaurs is the stegosaurus. Large—anywhere from three to nine meters in length and up to 6,500 kilograms in weight—the stegosaurus was a bit like a python on supports that has just swallowed a goat. It had a small head with a long thin neck, then it broadened right out with a massive belly carried by relatively short legs (longer at the back), and finally behind came a long, ever-thinner tail. Although very small-brained, the stegosaurus seems to have been a remarkably successful animal. It appeared in the mid-Jurassic (170 million years ago) and lasted at least until the late Cretaceous (70 million years ago). The number of species was not huge, but there were at least a dozen or more, and the stegosaurus lived all around the world—North America, Europe, India, and China. It was first discovered in the 1870s in England, but the dramatic finds came a few years later in North America—a time of fierce competition between the American paleontologists, Edward Drinker Cope and Othniel Charles Marsh. Complete specimens were unearthed from digs in Wyoming and elsewhere in the West. More recently, Asia has yielded the greatest number of specimens.

The stegosaurus was not particularly fast-moving—probably able to hit a top speed of about seven or eight kilometers an hour—but it would not have needed great abilities in that direction because it was a herbivore, foraging leaves and fruits and the like on or close to the ground. Given that its teeth were relatively small and its stomach relatively large, most of the fragmentation of food probably took place after it was swallowed. In some animals today—birds and crocodiles, for instance—stones are used by the stomach muscles to finish what the jaws and teeth left undone. Obviously, whatever the method used to cut up the fodder, it must have been fairly efficient given the brute's size.

We do not know much about the stegosaurus's home life. Some dinosaurs were highly social and exhibited fairly sophisticated behavior involving parental care, but the tiny brain of the stegosaurus (tiny not just in absolute terms but also with respect to the body that had to be moved about) suggests that any behavior was going to be fairly simple. Some evidence of sexual dimorphism exists, although no one seems to have much idea about which sex is associated with which set of bones and therefore which sex might be the larger.

The feature that made the stegosaurus truly remarkable was not its bulk or family life or diet. It was rather the series of bony plates that ran down the animal's spine from front to back, finishing with spikes at the end of the tail. Two major (interconnected) questions have been asked about these plates. First, how were they positioned on the animal? Second, what was their function? It is difficult to answer the first question because the plates were not connected directly to the backbone or other parts of the internal skeleton. They were embedded in the skin and other soft tissues. Some have suggested that the plates lay flat along the back, sticking outward. Others have suggested that the plates normally laid flat or were retracted and could be raised as needed. Recent studies have shown that almost certainly these suggestions are wrong and that the plates were upright and stayed that way. There is evidence of connecting fibers of the kind that, in today's organisms, connect bone to tissue (ligaments). In the stegosaurus, these fibers appear to have been located symmetrically on both sides of the bases of the plates, compatible with and only with continuous upright stance. Equally problematical is whether the plates were paired symmetrically down the spine or whether they were staggered, possibly overlapping. The plates are more generally found in a staggered position, and they do not match exactly in pairs, but these could be artifacts of fossilization. Most recent evidence does suggest staggering, and as we shall now see this ties in with answers to the second question about function.

Given their rather ferocious appearance, the obvious answer to the question of function is that the plates in some way protected the otherwise defenseless animal. This explanation was a major reason why some experts suggested that the plates might lie horizontally, to ward off predators. Plates that stick right up are less obviously helpful in that way to the animal, and now fairly definitive evidence suggests that the plates were not used in any direct fashion for fighting or protection. Detailed analysis of the bone making up the plates (one plate was cut into small pieces and its interior was studied with much care) shows that it was simply not the tough and strong kind of bone that is usually associated with fighting and combat. It tended to be fragile and fairly porous. It is possible that the plates were used to frighten would-be predators or that they were used for recognition of fellow species members or for sexual display. We do know that some dinosaurs had quite sophisticated recognition systems—the hadrosaurs, for instance, had a complex method of sound signaling, using air blown through their highly elaborate skulls. In the stegosaurus case, given that the plates were not retractable, the display option is somewhat less plausible. Things like the peacock's tail tend to be turned on or off as the need arises. Still, in the absence of much understanding of stegosaurus sexuality, no one can be very definitive on this matter.

But none of this really strikes at the heart of the matter. Fortunately, a much more convincing hypothesis about the primary function of the plates has emerged in the past few decades. Given the intense interest in the possibility that dinosaurs might have been hot-blooded, the suggestion that the plates were perhaps involved in heat regulation comes as no surprise; they would have been similar to the fins of generators in power stations, where such protrusions are an essential part of the heat-transfer mechanism. Could the plates have been used to cool off the dinosaur when it started to overheat—from, say, the huge number of calories generated by the fermentation process of its digestion? Two different approaches have been used to answer this question. First, scale models were built to see if and how such plates could transfer heat and if, in real life, they could be truly effective, given the known size of the animals. (No attempt was made to achieve exact verisimilitude. The models were aluminum cylinders with veins attached.) It was found that radiation would not be a very effective way of eliminating heat, but that wind-driven convection was a different matter. The models were placed in tunnels where such factors as temperature and wind velocity could be controlled and measured, and the results were very encouraging. Indeed, the fact that the plates might not be movable or retractable would have been no handicap, so long as the brutes could control the rates of flow of blood through the plates.

Particularly interesting was the fact that plates are more effective than continuous fins, and that staggered plates are more effective than matched plates. The full significance of this last point depends on whether or not—what now seems to be the case—there is independent evidence of the staggering.

The second approach involves close study of the structure of the plates themselves. Here also the evidence points to heat transfer. The plates have features that suggest that blood could be carried within the plate and moved close to its skin. (Although no skin is fossilized, given what we know about the susceptibility of exposed bone to bacterial infection, it is highly probable that there was skin and perhaps even a more horny type of covering overall.)

If heat transfer through convection is truly the adaptive function of the plates, then from this we can make a number of obvious inferences, particularly about the ecology of the stegosaurus. We can infer, first, that in order for the plates to function as a means of dispelling heat, the stegosaurus would have lived on open land, savannas, or similar grassy plains, with lots of wind of a fairly predictable kind. If the plates had a two-way function—not just dispelling heat but absorbing heat when the animal was cold—one would expect at least some regular patches or areas of open space, where the sun would shine through and heat up the animals. In other words, one would not expect the stegosaurus to live in a jungle or forest. As it turns out, the stegosaurus apparently did live on relatively open territory, where wind would be a significant factor in heat loss and where sunlight could help it warm up on chilly days. The fossil evidence suggests that this match between animal and habitat did not come about just by chance. Whereas many other dinosaurs of its generation seemed to be fairly indifferent to their surroundings, the stegosaurus apparently made special efforts to stay away from swampy, enclosed spaces.

Unanswered questions remain, including some fairly basic ones about whether indeed the plates might also have functioned to absorb heat when the animals were cold. If dinosaurs were not truly warm-blooded like mammals and birds, then methods of absorbing external heat could have been adaptively very significant. But empty spaces notwithstanding, overall, paleontologists feel that they now have a handle on one of the biggest and most interesting questions in their trade. A triumph of adaptationist thinking.

Infanticide

Darwin thought that one can apply natural selection to our own species, *Homo sapiens,* and later evolutionists agreed with him. In 1962 Theodosius Dobzhansky wrote an absolutely splendid overview of the (Darwinian) evolutionary biology of humankind. With the coming of an evolutionary approach to social behavior, one might have predicted extra effort in this direction. After all, we humans are the social species *par excellence.* And so it proved. Much has been written about human nature, in an attempt to tie it to our biology. Our sexuality, our family structures, our relationships with friends and enemies, our religious beliefs and rituals, our eating practices—all of these topics have been the subject of study and hypothesis. Like Darwin, William Hamilton thought that warfare was a crucial selective factor in human history. Like Darwin, Edward O. Wilson thinks that differences between women and men are a direct outcome of the workings of natural selection. Recently, Darwinians have been applying their theory to matters of health and medicine. Morning sickness, for instance: Is this just an unfortunate physiological by-product of pregnancy, or is it adaptive? Could it be that morning sickness protects the fetus at its most vulnerable stage of development from the possibly harmful effects of foodstuffs that are of no harm to adults?

Often, books and essays on the evolution of human social behavior have been speculative and controversial. So much so that today's enthusiasts—eager to escape both speculation and controversy—hide their activities under innocuous-sounding names, like evolutionary psychology. By way of illustration, I focus on a discussion by Sarah Blaffer Hrdy, an expert on primate and human social behavior. Hrdy is interested in a phenomenon that fascinated Darwin, along with many of his fellow Victorians: infanticide, the killing of babies or very small children by parents, usually the mother. The classic fictional case is Hetty in George Eliot's *Adam Bede.* At first appearance, such behavior seems very non-Darwinian. How can one survive and reproduce if one is killing off one's own offspring? An answer might be that a woman does so if for some reason her expectation of raising a healthy or successful child is reduced, perhaps because the child is handicapped or the mother is alone or without support. This was Hetty's situation precisely. In the real world, Canadian data show unambiguously that young, unmarried women are more prone to infanticide than women who are older and married—those who generally are in a better position to raise their children.

But what about more systematic infanticide? What about societies that accept killing or neglecting babies of one sex rather than the other, or even put pressure on parents to commit infanticide? To appreciate the problem, start with a question. Why should the sexes be of equal number? The argument to design had an answer. In the eighteenth century, Queen Anne's physician, Dr. John Arbuthnot, took such equality to be a sign of God's forethought and a mark of the propriety of monogamy. Realizing that males are more sickly than females, God even prepared for a higher mortality rate in males by giving them a slight edge in number at birth. Darwin realized that natural selection had to give its own answer, to combat the natural theologians, and he eventually found it. We can find that answer in the first edition of the *Descent of Man*; it dropped out of the second edition because by then Darwin was persuaded that he was wrong, and so he left the problem unsolved. In *The Genetical Theory of Natural Selection*, Fisher publicized Darwin's answer, which he rediscovered. These men showed that sex ratios are essentially a result of what biologists today would call an "evolutionarily stable strategy." If a species has a preponderance of one sex over the other, then it would be a good strategy for an organism to have offspring of the other sex. More opportunity, basically. If lots of girls are around, then it is better to be a boy, and conversely. This would be the case until an equal balance was achieved and then stability would ensue.

Here now comes the problem. In various parts of the world, we find that the ratios are way off the expected equilibrium. Notoriously in China, with the state's one-child-per-family policy, people feel enormous social pressure to have a son, and a consequence is widespread female infanticide (or, today, prenatal sex determination and abortion). The 1990 census showed that the male/female survival ratio was over 110/100. If the first child is a girl, then one can get a dispensation to have another child. The pressure to have sons rises rapidly, so that for families with five children (where presumably all of the first four children are female), the ratio of males to females among fifth-child survivors rises to 125/100. Why would this be so?

The China situation is by no means unique: in nineteenth-century India, for instance, in some groups absolutely no girls at all survived. The British enacted umpteen anti-infanticide laws, the main effect of which was wholesale forgery of the census returns. Daughters continued, as is indeed still often the case, to have a hard time making it, whereas sons were treasured and well cared for.

Various kinds of cultural explanations could be given, and some are extremely plausible. A Chinese farmer, for instance, needs some kind of insurance for old age, and a son working the family farm is a better bet than a daughter, who would marry into someone else's family and contribute to its well-being. In India, daughters even had to bring dowries to their new husbands, which added to their expense. Darwin pointed out that if a tribe was under threat from neighbors, sons would count higher as possible defenders than would daughters. So, from the beginning, no one has denied that whatever explanation the Darwinian offers must posit causes that work *with* culture rather than against it.

A Darwinian theorem that does just that was proposed in the early 1970s by Robert Trivers and Dan Willard. They argued that, given reason to think that there will be significant differences in the chances of reproductive success between males and females, then this should be reflected in differences in sex ratio and should be tied to the status of parent or parents. Suppose, to take the usual pattern, most females are likely to become impregnated and give birth no matter what, whereas most males are going to have to compete for mates, with some males being successful (perhaps very successful) and others being unsuccessful (perhaps totally unsuccessful). Then one would expect those parents who are able to give a high level of care to favor male offspring, gambling that their sons will be reproductively successful, and one would expect those at the other end of the spectrum to favor female offspring, realizing that sons would have little chance of reproductive success.

The Trivers-Willard hypothesis holds in the animal world. An extensive study of coypus—the South American guinea-pig-like creatures that were imported to Britain and became feral—showed that the fattest females were giving birth to far more males than females; when they were carrying litters of mostly females, they spontaneously aborted. The underfed females, to the contrary, carried mostly or exclusively female litters to term. Somehow, the biology of a coypu female assesses its own condition and aborts or not, depending upon its best reproductive interests. This is not an isolated case. The hypothesis holds true of many, many species in the animal world. High-ranking females deliver male offspring. Low-ranking females deliver female offspring. Low-ranking female spider monkeys in Peruvian rain forests rear females almost exclusively.

But what about humans? The anthropologist Mildred Dickemann reexamined the data from India where female offspring seemed to be missing or well below equilibrium and found, precisely as Trivers and Willard predicted, that high-status families specialized in males. Lower-status families were much more likely to keep their female offspring and let their sons fend for themselves. These lower-status families were saving up dowries for their daughters, who would then marry into higher-status families, ensuring that a good social connection was made (useful in case of family disaster) and that the surviving offspring (the grandsons, that is) were themselves of higher status. Technically, high-class females were allowed to survive and to marry, but they rarely or never did. This hypergamy is consistent with the Trivers-Willard hypothesis, which predicted a "tendency for the female to marry a male whose socioeconomic status is higher than hers" (Trivers and Willard 1973, 91).

Genetic evidence suggests that hypergamy is a long-standing practice in India, though it is far from unique there; Hungarian gypsies also tend to favor daughters. Humans follow Darwinian lines for adaptive reasons. It is not a matter of genes versus culture but rather a case of genes getting their way through culture. Which raises the final question of why humans do it deliberately rather than by purely physiological means, as other animals do? Why do we not just spontaneously abort when the desired sex is not being produced? In fact, humans are not entirely distinct in this respect. Other animals do sometimes manipulate the survival of offspring after birth. Male lemmings, for instance, kill the kits of another male when they take over a new female.

In the human case, the answer is partly that we can manipulate ratios through culture and partly that we need to do it through culture. Because of our intellectual abilities, we can change the situation more readily after birth and so we do so. On the other side, we live in an environment where pressures can change rapidly, through outside natural forces or human-inflected ones. Major biological value accrues to those who can assess the situation right up to the last moment. A drought or a conquering raid could change one's prospects literally overnight. Hence, far from our culture making us non-Darwinian, in a way it makes us super-Darwinian.

But culture also allows us to escape Darwinism to some extent. In societies such as those of the West, with their vast material wealth, the pressures to control offspring are not nearly so strong, and male and female offspring tend to be evenly cherished. Before we become too condescending about other cultures, however, we should remember the longstanding tradition that assigned European daughters of nobility to the equivalent of female infanticide, namely, to the nunnery and a lifelong vow of chastity. Of course becoming the bride of Christ could be seen as the ultimate act of social climbing.

Aristotle warned: one swallow does not make a summer. Five case studies can only illustrate selection at work, not prove it definitively. But that is goal enough for now. In the realm of the argument to complexity, Darwinism thrives. Natural selection is a powerful tool of adaptation, which is widespread in the living world. Let us now take this conclusion and turn to critical discussion.

CHAPTER 10

LET US BEGIN with the most basic question. If one is a Darwinian—meaning someone who takes natural selection as the absolutely fundamental mechanism of evolutionary change—then in practice or in theory can one allow the existence of the nonadaptive? Can a characteristic survive that has no value at all, relative or absolute? Can it even detract from value and still be retained in later generations? Or is natural selection a tautology, as critics often claim? That is, those that survive are those that survive, and no conclusions can be drawn about biological value.

At one level, this is an easy question to answer. If only because of time lags, organisms are sometimes poorly adapted to their environment. Consider the English cuckoo. This bird lays its eggs in the nests of birds of other species and then takes off for its wintering grounds in North Africa, leaving the hosts to raise its chicks. The cuckoo nestlings have adaptations for ejecting the eggs and chicks of the host, so that they alone remain in the nests and monopolize the attentions of their foster parents. These chicks need exclusive parental rearing because they tend to be very much bigger (and hungrier) than the hosts' own chicks or even the host parents themselves.

As you might imagine, a consequence of all of this is a fierce selective contest between cuckoos and hosts. Cuckoos have all sorts of special adaptations to aid their parasitic behavior. For example, whereas most birds like to take their time laying eggs, the female cuckoo is able to drop an egg at lightning speed whenever the opportunity arises, thus allaying suspicions of the potential hosts. Conversely, the parasitized birds become very adept at spotting strange eggs and ejecting them or abandoning the nest at any hint of a stranger in their midst. And as expected, the cuckoos have counter-attacked, producing eggs that are nigh identical to those of the host, in size, shape, and color. None of this is mere chance, for different cuckoo strains specialize in different hosts, and the eggs perfectly imitate those of the respective host in question.

Except for our feathered friends, the dunnocks! They too are parasitized by cuckoos, but cuckoos make no attempt to disguise their eggs, nor do the dunnocks discriminate against them. This point was noted in the eighteenth century by the Reverend Gilbert White, for he wrote in his *Natural History of Selbourne* (letter V to Daines Barrington): "You wonder, with good reason, that the hedge sparrows can be induced at all to sit on the egg of the cuckoo without being scandalized at the vast disproportioned size of the supposititious egg; but the brute creation, I suppose, have very little idea of size, colour or number" (Davies 1992, 218).

According to Nicholas Davies, we have here a clear case of adaptive failure, a conclusion reinforced by the experimental fact that when eggs are shifted around in host nests (including artificial or "model" eggs), all the species he studied are highly sensitive to the changes, except the dunnocks. They simply are fooled, despite the fact that the cuckoos make no attempt to deceive them.

Maladaptation

Design has broken right down in the dunnock world. Why? There are probably complex reasons, but the most obvious hypothesis is that dunnocks have only recently been parasitized by cuckoos, and they have not yet developed adaptive defenses. A side experiment, described by Davies, was very suggestive. In his own words (pp. 229–230):

> The cuckoo breeds from Western Europe to Japan but not in Iceland, where it is only a rare vagrant and has never been known to breed. Iceland does, however, have isolated populations of meadow pipits and white wagtails (of which the pied wagtail is a subspecies). We therefore took our model eggs to Iceland. The Icelandic populations bred at low densities and we had to work very hard to find nests but the results were exciting. Both the pipits and wagtails showed much less discrimination against eggs unlike their own than did members of the parasitised populations of these two species in Britain.

The case is not closed absolutely, but the point is firm: even the most ardent adaptationists recognize gaps and failures.

What about more systematic failures of adaptation and design? At some basic level, this comes with the territory, for it has always been emphasized, from Darwin on, that the building blocks of evolution—the raw variations—do not come (as Asa Gray supposed) according to need but purely by chance, in that they are not designed for immediate use. Hence, in some sense adaptation is always a struggle, no matter how efficient the power of selection. If the desired material is never there, you cannot cut your cloth to taste—you may not have any cloth to cut at all. But for Darwinians, this fact of nature is not terribly constraining. Fisher in particular thought that the randomness of mutation meant little to the course of adaptation and evolution in the long run.

Take size and color as examples. Granted, if the mutations do not appear, then certain moves are simply impossible. However desirable an increase in size or change of color might be, if the variation is not there, then it is not there. But remember that organisms are package deals and that macro-qualities like size and color are rarely caused by one and only one gene or set of genes (alleles). Hence, that no variation in such qualities will be possible or available is highly unlikely. Also, for much of the time populations have a bank or library of variation encoded in their genes and available for use as the need arises. One should therefore not think that the nondirectedness of mutation will always cash out as maladaptation. Bates's mimicking butterflies, to take but one example, seemed to have had little difficulty in picking colors and patterns from their palettes at will.

Genetic Drift

More interesting than maladaptation, and perhaps more significant although certainly more controversial, is genetic drift (the Sewall Wright effect). According to this theory, the smallness of a population means that random mating can cause permanent changes on its own, independent of natural selection pressures that might be pushing the organism in another direction. In the 1940s and 1950s, the popularity of genetic drift among evolutionists, especially in Europe, dropped like a stone, but in America a fondness for the concept still lingered. Admittedly, many of the supporting cases did not hold water—including Wright's own favorite of the desert plant *Linanthus parryae.* But some evidence for the existence of genetic drift still seemed convincing. For example, caged populations of fruit flies were shown—over generations—to vary in ways that one would predict from drift. The same is true of some natural populations.

In the 1950s Ernst Mayr introduced a significant corollary to genetic drift, the so-called founder principle. Mayr claimed that when a small group of organisms becomes separated from its parent population, through interbreeding it can generate dramatic physical changes in subsequent generations and eventually become a separate species from the parent group. This happens because small founder populations differ genetically from others in the parent population, owing to the variation in all natural populations. Again, evolutionists have argued about how prevalent this phenomenon truly is, but some evidence of its existence is widely accepted. We know, for example, that species of presumed recent vintage have reduced variability in their genomes, and this is what we would expect if these species originated as small founder populations.

The founder principle may offer insight into the evolution of humans. If, as many suppose, most humans stem from small founder populations that migrated out of Africa some 150,000 years ago, then you would expect more genetic variation in the parent stay-at-home groups and less in the wanderers. And indeed this is the case: native African populations show more genetic variation than do people native to other continents.

One hypothesis about genetic drift has attracted considerable interest in the past half century. This is the theory of molecular drift (also known as the neutral theory of molecular evolution), proposed by the Japanese theoretical population geneticist Moto Kimura. He agreed that at the normal physical (phenotypic) level, natural selection works and rules, but he argued that at the level of DNA much of the variation one sees is random: it is not exposed to selective forces in the environment and hence drifts without aim or purpose. The plausibility of this hypothesis is supported by a number of phenomena at the molecular level. First is the fact that much DNA does indeed seem to have little purpose; it is called "junk DNA" because it does not encode for proteins or serve any regulatory function that we know of. This seemingly superfluous DNA would be a prime candidate for drift. Second, the very way in which the DNA molecule yields up its information seems to invite drift. DNA makes RNA, which then is "read" to make amino acids, which are in turn linked into long polypeptide chains (proteins). DNA is a chain of linked bases, of which there are only four types, and the same is true of RNA. However, there are twenty amino acids. This means that the minimum genetic code conveying information from DNA to amino acid must involve three bases (yielding 4^3 codons), for any less would not distinguish all of the different amino acids (since 4^2 is less than twenty). But this also implies considerable redundancy in the codons (since 64 is much bigger than 20). And this is in fact what we find: the third base in particular is often insignificant in defining an amino acid. Even if selection were operating flat out, it would seem that in such cases the third base would drift aimlessly.

That molecular drift is important seems agreed upon by all. It is the basis of so-called molecular clocks, where the time since important speciation events occurred can be estimated by measuring the genetic difference between groups—the greater the difference, the greater the time since splitting. Over time, however, random drift would be expected to average out more smoothly than differences due to the ever-changing forces of selection. For this reason the hypothesis that most molecular difference is due to drift has not been well received. Time and again, measurements have shown that molecular differences are not what we would expect were drift the sole or main cause of change. In fruit flies, we saw how random drift was ruled out as a significant factor in changing levels of the *Adh* gene.

In the next chapter, we will look at some quite remarkable instances of gene sequences being conserved for hundreds of millions of years, in the face of mutation and drift and any other degenerating factors. Also, although there may be room for wiggle in the third base of a codon, it does not follow that selection is entirely indifferent. Although two sequences may code for the same amino acid, one sequence might code more efficiently or quickly and hence be favored by selection. And in fact considerable evidence of "codon bias" has been uncovered, suggesting that although drift may count at the genetic level, so also does selection.

When we turn later to matters of form over function, we shall encounter other cases of (putative or supposed) nonadaptation. Although evolutionists fiercely debate the extent and importance of this concept, no one denies that some aspects of the organic world are not well adapted, if they are adapted at all.

Perfection

Turn the question around the other way. Is adaptation always striving toward perfection—a perfection that might never be achieved but, like the Holy Grail, is always there to be sought? That adaptation is frequently not perfect—that it fails by obvious design standards—is agreed by all. Consider the urogenital system of the human male. Thanks to our evolutionary history, the sperm duct is rather like a garden hose that takes an unneeded loop around a distant tree, on its way from the tap to the nearby flower bed. The duct got itself hung over the ureter, and so instead of going directly from testis to penis, it meanders around. Hardly a triumph of plumbing, although nothing about any of this surprises evolutionists, and certainly nothing about it threatens Darwinism or adaptationism as such. Anything but. The urogenital system is good evidence that evolution has been at work here rather than a grand, directly intervening intelligence. If, as evolutionists believe, organisms are built one step at a time, always working from what they actually have rather than what they might like to have, then we would expect these kinds of jerry-built effects. No one is saying that the urogenital system is not adaptive or that we would be better off without it. The point simply is that we must think in a relativistic sense—is it better than its competitors?—rather than an absolutist sense—is it the best that could ever be imagined?

Relativism is a factor even when selection is working well. At the individual level, adaptive excellence in one domain often has to compromise with adaptive failure in another domain. Many genes have pleiotropic effects, meaning that they produce the proteins for more than one physical feature. Selection may well favor one of these features more and the other less or not at all. Or perhaps selection favors a feature at one stage of life but not at another. Allometric growth—where a feature grows at a rate different from the rest of the body—may be a key factor in adaptation. It is possible that the Irish elk, notable for its too-magnificent antlers, fell victim here. Selection favored rapid development and large antlers in adolescents, so that they could start breeding early and efficiently. Unfortunately, the animals went on growing and the antlers even more so, until they became positively disadvantageous. All-around perfection was simply impossible in such a situation. Sexual selection and natural selection were working toward different ends.

As we now start to move toward the group level, we should note that natural selection may promote diversity rather than uniform perfection. Some of Bates's butterflies mimicked one model and other forms mimicked other models. From the standpoint of survival, this variation rather than some absolute mode of perfection was a huge advantage. In some species there may even be selection for rareness. If mimics get too popular, the predators will start to learn that they are being fooled and adjust their own survival strategy. And even in situations where mimicry is not a factor, if predators survive by learning to recognize their prey quickly, individuals with an uncommon appearance may have an adaptive advantage and go undetected as dinner. Unfortunately such an advantage has a feedback consequence: the once-uncommon appearance becomes more common in the population as a whole, so that the better it works the sooner it brings on its own failure.

As behavioral biologists have turned to game theory to help them explain selection, they have placed even more emphasis on selection-maintained diversity in populations. Some of Davies's male dunnocks are alphas and some are betas, all trying to maximize their reproductive opportunities in the face of competition from others. Perhaps all would like to be alphas, but this is impossible so it is better to be a beta than nothing at all. The point is that no script requires that selection lead to uniformity or that all adaptation go in an identical direction. The thinking, indeed, is much the other way around.

Let's move now from questions about excellence today to questions about excellence over time—to questions about so-called macroevolution. Do we see that adaptation improves the living world in some general way over the ages? Some seem to think that orthodox Darwinians do or should see much order and control in the overall pattern of evolution.

Evolution should show long-term trends in forms, these trends clearly being powered by selection and resulting in improved adaptations. In the long view, one should see progress, from the small and simple to the large and complex, from the "monad" to the "man."

As it happens, many Darwinians would protest this conclusion, but let us start with some areas of agreement. First, that trends exist seems to be uncontested. When a radically new form of animal emerges, it often starts out rather small, at least compared with the size of its descendants. The horse, for instance, grew mightily from little Eohippus, smaller than a Shetland pony, to today's hearty animals, bigger than people. Likewise, the human line shows increase in size over time, from the three-foot-plus Lucy (*Australopithecus afarensis*) to the five-foot-plus modern human (*Homo sapiens sapiens*). It could of course be that the size increase (known as Cope's Rule) is more—or at least in part—a factor of the smallness of the initiators (linked to the rapid change to a new form) than the largeness of the descendants. But fairly uncomplicated selective factors behind the changes are not so hard to imagine. Especially with respect to one's competitors, greater size has its obvious adaptive advantages.

In fact, when faced with trends such as increased size, Darwinians today usually trot out the notion of arms races. Regarding the evolution of the ungulates, Julian Huxley (Thomas Henry's grandson and himself an evolutionary biologist) wrote (1942, 495–496):

> It did not take place in a biological vacuum, but in a world inhabited, *inter alia*, by carnivores. Accordingly, a large part of ungulate adaptation is relative to the fact of carnivorous enemies. This applies to their speed, and, in the case of the ruminants, to the elaborate arrangements for chewing the cud, permitting the food to be bolted in haste and chewed at leisure in safety. The relation between predator and prey in evolution is somewhat like that between methods of attack and defence in the evolution of war.

More recently, the invertebrate paleontologist Geerat Vermeij has written extensively of the ways in which shells are strengthened against predators, which in turn develop ever-stronger and more sophisticated ways of breaking into those fortified shells. These sorts of ideas have not gone unchallenged. On the basis of the fossil evidence, it has been argued that the predator/prey increase-in-speed hypothesis does not find universal confirmation. Many cases have been found where a maximum speed seems to have been reached and stabilized over long periods of time. So although arms races may be important, they are probably not all-important.

What about bigger patterns? Do we see overall levels of improvement? Do we see evolutionary progress? Some evolutionists have good reasons to think so. Even catastrophic events that cause mass extinction do not stop—and perhaps even aid—progress. Take the asteroid or comet that hit the earth and brought about the demise of the dinosaurs. This was not by any means the first mass extinction, or even the greatest one—although whether extraplanetary bodies were responsible for other extinctions or whether other causes, such as the shifting of the continents, explain them is still a matter of debate. It is true that mass extinctions did not completely shift the course of history, as far as we know. No new phyla were created by them, for instance. And in all cases, some organisms survived the event and diversified on the other side, in the new ecological landscape. But extinctions did make a major difference. Had the dinosaurs not died off, the mammals—small nocturnal creatures that had already been around 150 million years, almost as long as the dinosaurs, without making a major breakthrough—might never have taken over the world. As Stephen Jay Gould said: "Since dinosaurs were not moving toward markedly larger brains, and since such a prospect may lie outside the capabilities of reptilian design . . . we must assume that consciousness would not have evolved on our planet if a cosmic catastrophe had not claimed the dinosaurs as victims. In an entirely literal sense, we owe our existence, as large and reasoning mammals, to our lucky stars" (Gould 1989, 318).

Gould in fact was at the forefront of those who argued that the pattern of evolution in larger groups (at the taxonomic level of families) shows no evidence of selection, with diversity within these groups being no more than what one might expect by chance. However, one of Gould's co-authors, the paleontologist Jack Sepkoski, qualified and modified this position. Looking at marine invertebrate families since the Cambrian (over 500 million years ago), he argued that one sees a definite, repeated picture, with life forms exploding in number and then stabilizing (or starting to decline) and new explosions of life forms then building on what has gone before. Sepkoski was more interested in the dynamics of the picture than in the actual causes. But, as he suggested, one assumes that the explosions had something to do with empty ecological niches, and that life-form explosions tapered off as spaces were filled. One would expect also to see (as apparently one does see) a move toward more specialized organisms, reflecting the refinement of adaptations as the going gets tougher. New explosions would represent the move to radically new life forms. But again, as with size, being specialized rather than generalized is not necessarily better in some absolute sense, even though in competition a specialist may usually beat a generalist. So even at the macro-level, some improvement is probably more relative than absolute.

This reflection raises the final and biggest question of them all. Is there nevertheless some overall pattern to the history of life? Do we start with the simple and work up to the most complex and important, *Homo sapiens?* This was certainly Darwin's view. Right from the beginning, he was insistent that while there could be no necessary progress up to humankind, the sheer pressure of trying to find new ecological niches would force organisms to become ever-more complex. "The enormous number of animals in the world depends, on their varied structure & complexity.—hence as the forms became complicated, they opened fresh means of adding to their complexity" (Notebook E, 95). All of those arms races ultimately cashed out in terms of the triumph of humankind. In Darwin's words: "If we take as the standard of high organisation, the amount of differentiation and specialisation of the several organs in each being when adult (and this will include the advancement of the brain for intellectual purposes), natural selection clearly leads towards this standard: for all physiologists admit that the specialisation of organs, inasmuch as in this state they perform their functions better, is an advantage to each being; and hence the accumulation of variations tending towards specialisation is within the scope of natural selection" (Darwin 1959, 222).

Today, according to Richard Dawkins, arms races in nature have gone the way of arms races in the military, from clashes between armor plate and guns to the development of sophisticated electronic methods of attack and defense, from clashes between shells and boring mechanisms to the development of ever more sophisticated on-board computers, otherwise known as brains. All of which led the entomologist and sociobiologist Edward O. Wilson to write: "The overall average across the history of life has moved from the simple and few to the more complex and numerous. During the past billion years, animals as a whole evolved upward in body size, feeding and defensive techniques, brain and behavioral complexity, social organization, and precision of environmental control—in each case farther from the nonliving state than their simpler antecedents did" (Wilson 1992, 187). He saw but one conclusion: "Progress, then, is a property of the evolution of life as a whole by almost any conceivable intuitive standard, including the acquisition of goals and intentions in the behavior of animals."

Others will have none of this. For Gould, progress was "a noxious, culturally embedded, untestable, nonoperational, intractable idea that must be replaced if we wish to understand the patterns of history" (1988, 319). We put ourselves on a pedestal, read this into biological history, and declare ourselves the winners. Natural selection could hardly be the cause of progress, for there is no progress. But is this so? Part of the problem here is defining precisely what one means by "progress" in the biological context. What one is usually after is something that puts humans on top, but one cannot simply specify "human-like" as the desired quality, for that simply makes progress true by definition.

Still, by some measures that seem intuitively plausible we might reach the conclusion that some kind of progress has taken place in the course of evolution. For instance, over time the length of a species' life before it becomes extinct seems to have increased overall, suggesting that perhaps adaptations have become more sophisticated. On the other hand, other measures that attempt, for instance, to capture some notion of complexity suggest that not much progress has occurred. The back of the whale is less complex than that of its dog-like land ancestor, and yet would one want to say that the whale is less advanced than a land animal? (Or, for that matter, that it is more advanced?) And even if one agrees that something important has occurred—and, whether by definition or not, it is hard for us humans not to think that something important has occurred—this "something" could be more a function of a kind of drift than anything positively fueled by selection. However random the process, on average organisms seem to be able to evolve only in the direction of more complexity, not less. It is like the drunkard on a sidewalk bounded by a wall on one side and the gutter on the other. The drunkard can never go through the wall but eventually his random stagger will land him in the gutter. Perhaps that is all one sees in the course of history—a random stagger toward complexity.

Clearly, questions about improvement are themselves complex and still fodder for debate. No one wants to say that selection always leads to improvement, and no one wants to say that there is always one unambiguous standard of excellence. At least, no one speaking purely as a scientist will want to say it. But clearly here we have a subtext about values. To suggest that the evolutionary process has one aim, total perfection, implies the sort of absolute values that the Darwinian revolution expelled from science. At best one can say that evolution leads to that which we humans value most of all, namely ourselves. But the value has been read into the text, rather than discovered in it. And in any case, selection itself often points away from unique measures of perfect adaptation. The fairest conclusion, therefore, is that, with respect to the question of progress, people are very much divided.

Levels of Selection

In the last chapter we encountered William Hamilton's idea of kin selection, where adaptations that apparently benefit the group can be shown to be the result of selection for the genetic success of the individual. This mechanism points to a broader issue—one that we have been edging toward at times in the discussion above—an issue that has been much debated in the past three or four decades. If we grant kin selection, and grant all of the other mechanisms that are invoked by students of social behavior like Davies, is this as far as we can go? Have we reached the end of the line on the subject of levels of selection? Can one truly say that selection works only at one level and that adaptations are never selected for any purpose other than aiding the genetic success of the individual organism? That any benefits for other organisms, including the group as a whole, are tangential at best? And what about the way in which Darwinians like Hamilton seem to jump casually from talk about organisms to talk about the selective virtues of various genetic configurations? Even if we allow selection on and for the individual organism, how can we then at once flip into talk about selection on and for the individual gene? Many commentators have found something deeply problematical in talk of "selfish genes." Even if we grant that the concept of selfishness is metaphorical, selection generally works on whole organisms, not genes, and adaptation is usually something that exists at the macro-level of the senses rather than the micro-level of the cell. Talk of selfish genes is surely pushing reductionism—explaining the smaller in terms of the larger—to the extreme.

To disentangle and explore some of these issues, let us start with the fact that, whatever the action of natural selection, in a very important way the genes are indeed at the center of things. Unless one rejects completely the approach taken since the work of Fisher and the other population geneticists, evolution is virtually defined as changes of gene frequencies. No gene change, no evolution. So if nothing else, the genes are used as counters or markers. More than this, though, since the coming of population genetics, changes in gene numbers and ratios have been assessed in causal terms, namely mutation, selection, immigration, emigration—and of course drift. None of this is so very different from the ways in which we might talk of gases in terms of molecules and of their velocities and impacts and so forth. So talk of selfish genes, in reference to the way that selection works on genes or because of them, may be rhetorically dramatic but it is hardly all that new or peculiar.

And in any case, although in a sense the modern Darwinian thinks that selection ultimately comes down to counting the proportion of DNA copies that gets poured into the next generation, this in no way means that the whole organism drops out of sight. A useful distinction is between the replicators, the units of heredity that get passed on down through time and record the changes in evolution, and the vehicles or interactors that package and carry the genes from one generation to the next. Granted that one may not always have full information (especially about the nature and workings of the genes), we can still say confidently that both levels—replicators and vehicles—are presupposed in a full understanding of the workings of natural selection.

This works fine for individual selection, and if you think something like kin selection is a fancy kind of individual selection—meaning that it is the success of the individual's genes that count—then we have pretty much covered and explained all that we have encountered so far. But what about group selection? Why was this concept so summarily dismissed by Darwin and his followers? No one has ever argued that group selection—meaning now a kind of selection that not only favors the group but favors the group at the biological expense of the individual—is conceptually impossible. It is just that they do not see how, other than in very rare cases, it could ever take effect. No matter how good a variation may be for the group, if it comes at the expense of the individual, the individual (the vehicle) who bears the genes favoring that adaptation will not survive to reproduce, and the genes will not be passed along to the group. And the genes of other individuals, whose different traits make them better able to survive and reproduce, will pass their genes along instead. In other words, individual selection will start to corrode group selection.

People like the group-selectionist Wynne-Edwards supposed that members of flocks of birds restricted their breeding practices because then the flocks would benefit from the extra room and available food. But as any game theorist would quickly point out, any individual who cheated, by having lots of offspring while enjoying the benefits that accrue from the rest of the group's restraint, would at once start to spread in numbers and wipe out the altruists making up the rest of the group.

So we must grant that this old-fashioned, uncritical group selection is as dead as the dodo. It was just plain wrong. In Fisher's words, "Natural selection would not explain any gentlemen's agreement among dogs not to eat each other" (Bennett 1983, 231–232). But even if group selection is generally without an empirical base, are the allowed rare exceptions really so very rare? Has a worthwhile baby been thrown out with some very dirty bathwater? Many today think that this is indeed so, although to a certain extent the discussion goes at cross-purposes. Although certain mathematical techniques can treat aspects of selection in terms of traits of groups rather than just individuals, and although the term "group selection" is sometimes used in these contexts, this does not constitute a resurrection of the kind of group thinking that was rejected by evolutionists in the 1960s.

More interesting than merely relabeling examples of individual selection as group selection is the concept that whole organisms are not necessarily the only entities that should qualify as vehicles. Indeed Darwinians argue at great length about what precisely constitutes an organism. Edward O Wilson, for instance, thinks that some species of ant have gone so far down the route of sociality that now (as Darwin argued) the whole colony rather than separate ants should be considered the unit on which selection acts. Perhaps things also change where culture is involved. For all of his stand against group selection, in the *Descent of Man* Darwin himself weakened and supposed that perhaps an individual might sacrifice himself for the good of the tribe, although probably for Darwin the tribe members were interrelated, so a kind of kin selection was at work. And as soon as he floated this idea, Darwin reverted to an individualistic stance, suggesting what today is known as reciprocal altruism: "As the reasoning powers and foresight of the members became improved, each man would soon learn that if he aided his fellow-men, he would commonly receive aid in return" (Darwin 1871, 1.163).

After possible exceptions like these, controversy really starts to boil when other candidates are proposed. For effective group selection, you need extreme circumstances, where groups are very strongly favored and where a group force can take effect before the reproductive success of individuals can erode it. The English evolutionist John Maynard Smith (1978a), whose claim to fame rests on his penetrating use of game theory to solve evolutionary problems—an approach that is the epitome of the new individualistic selection-based biology—allows (as did Fisher) that there might be one very important breach of the group selection ban. This is the case of sexuality. Most people think that sex is a very good thing for the group. A new advantageous mutation can be spread rapidly through the group through sexual behavior. But when you think about it, sex is a very bad thing for females. For all that exceptions (like birds) exist, generally, all females get is a squirt of sperm and the effort of raising a baby alone. They would be far better off if they just produced virgin females like themselves, asexually. Then they would be reproducing all of their own genes, rather than just half of their own and half of a stranger's.

Maynard Smith points out that strong selective forces must have maintained sexuality, for when you find related species where one is sexual and the other is not, the sexual species does as well as the asexual species. This force, Maynard Smith concludes, must be a group force, something that makes possible the rapid spread of new mutations that are advantageous for the group as a whole. Not surprisingly, not everyone agrees with him. Using the arms race idea, Hamilton for one proposed an individual-based explanation of sexuality. He pointed out that, thanks to the randomness introduced by the Mendelian laws, sexual organisms are constantly rearranging their personal gene ratios; other than identical twins, no two sexual individuals are the same genetically, even when they have the same parents. This means that, in the ongoing battle between hosts and parasites, sexual organisms present a constantly moving target to viruses and other intruders. Although the parasites are able to evolve much more rapidly than their hosts, the hosts stay one step ahead by constantly reshuffling their genes. Organisms that produce virtual copies of themselves through asexual reproduction would be much more vulnerable to a microbe invasion that would wipe out the entire population.

Do other levels of selection exist, above the population or species? Gould and other paleontologists have championed "species selection." According to this rubric, a species has a unity and integration unto itself that merits the use of the Darwinian mechanism of differential reproduction. In fact, species selection can just be a relabeling of individual selection—or more generally one can say that "clade selection" (a clade is all of the descendants of a species) is a relabeling of individual selection—and George Williams for one is happy to do this. If the members of one species of organisms are generally superior to the members of another species, then the first species might wipe out the second. One could call this species selection, although not in any sense that would challenge individual selection or boost group selection.

More intriguing is the possibility that species have "emergent" properties—properties of the species as a whole, not of the individuals that make up the species—that might be adaptive. One species property would be reproductive isolation. This terms refers to barriers—physical, physiological, or psychological—that prevent members of different species from interbreeding. Let's say that two species, A and B, are facing challenges from predators. One species puts its efforts into rapid multiplication of offspring. The other species puts its efforts into dispersion. In itself, neither strategy is superior to the other. However, if Mayr's founder principle is operating, then the second species will be much more likely to speciate—break up into more than one species—than the first species. Let us say that species A breaks up into C and D, but species B breaks up into E, F, G, and H. The daughter species have many of the parents' qualities, and again each is equally good in its way against the predator. But of course, now the daughter species of B are likely to have more variation overall than the daughters of A because more species have undergone revolutionary rearrangements of their genome. And this variation could be a decisive factor, even if the overall numbers of individuals in the two daughter lines are kept constant. In fact, even if there were no differences in the amount of variation in the new species, more species might survive in one line than in the other. One line could end up with just species C, while the other line has species E and F, say. And even if B's descendants were more likely to go extinct than A's descendants, so long as B was speciating much more rapidly than A, more B descendant species might survive than A descendant species. Here then is a kind of differential reproduction, one that Gould and his colleagues called species selection.

Apart from questions about how often this occurs, and even more questions about how important it is when it does occur, the pertinent question for us is what this all means in terms of adaptation and organized complexity. Group selection may never work, but there is no ambiguity about what it produces if it does work: benefits to the group at the expense of the individual. Species selection, by contrast, is not as clear on these matters. In the example just given, the speciation occurred by accident and its effects were not produced by selection, except insofar as selection in the form of prey pressure brought on wandering, which in turn brought on speciation, which in turn brought on genetic revision, which in turn brought on reproductive isolation. To use terminology introduced by the philosopher Elliott Sober, we may have selection *of* certain things, but I am not sure that we have selection *for* certain things. Even if some species make it where others do not, they were hardly produced to this end or cherished because of this end. The kind of circular process spotted by Kant—the tree produces the leaf which is needed by the tree (and leads to more trees)—is missing.

One cannot therefore really think of reproductive isolation as an adaptation. Nothing designed it. It just happened. Gould himself seemed to concede this. "A stronger link may exist between adaptation and selection at the organism level than at higher levels. This might constitute a genuine and interesting difference between these levels" (Lloyd and Gould 1993, 597). It might indeed. Of course, it might show that concerns about the lack of species-level adaptation is "inspired by misplaced emphasis on adaptation" (Gould and Lloyd 1999, 11908). But we can leave worries like this for the next chapter, when the question of adaptive emphases, placed and misplaced, will get a full airing. For now, we must turn from theory to testing.

Comparative Studies

The practical life of a Darwinian—in the laboratory or the field—involves trying to see how and where natural selection brings on adaptation, or, if it fails, why precisely these exceptions occurred. Our survey in the last chapter demonstrated a range of methods and tools available to the researcher, including the most recent findings and techniques of molecular biology. Ideally, in our studies of the intersection of natural selection and adaptation we would approach the kind of understanding now found in the study of the *Adh* gene in fruit flies. We would uncover the underlying genetic basis, find how the genes translate into physical characteristics, identify the selective forces, map the distributions, discover the variants, and more.

This is the ideal, obviously only rarely achieved in full. Much of the time we have to rest with less information, and clearly in many cases the ideal will be impossible to achieve. Despite the fantasies of *Jurassic Park,* the genetic structure of the dinosaurs is lost forever; no one will ever be able to run selection experiments on stegosaurus to see how the plates respond to different environmental pressures. Beyond the obvious approaches of observation and experiment, are there other tools that can come to the aid of the inquiring evolutionist? A comparative analysis may help. We can try to see if some characteristic (y) changes values in a systematic way as some other factor (x) changes its values. If we find a connection, then we may conclude that there is an adaptive link.

One of the more entertaining examples of recent years is a study by Oxford biologist Paul Harvey and colleagues on testis size in primates. The chimpanzee is incredibly well endowed, to put the matter bluntly. Its close cousin, the gorilla, is not—a shrinkage that becomes even more apparent when one factors in the overall sizes of the animals (gorillas are very much larger than chimps). A major clue to the cause of the difference is the sex life of gorillas, which is fairly temperate. A troop consists of a single male mated to one or more females, with their juveniles and infants. In a chimpanzee troop, which consists of several adult males, a number of females, and a host of youngsters and infants, the sex life of males makes even dunnocks look celibate. When a female comes into heat, the males of the troop engage her in multiple copulations, with a sexual frenzy unknown outside an Italian skin flick. Harvey therefore floated the hypothesis that among chimps there is significant competition among sperm after copulation, and the male who leaves the most sperm is at a distinct selective advantage. Hence the need for large and efficient testes in the chimpanzee but not in the gorilla, where the dominant male has exclusive and leisurely access to his females.

To test this hypothesis, Harvey and his associates did a large comparative study across all primate species, garnering information on testis size in relation to overall body size, and also on the social structure of species and their known mating practices. A very definite pattern emerged. Those species whose mating practices suggested that sperm competition might be a factor in reproductive success tended almost universally to have larger testes than the average species, whereas those species whose mating practices suggested little or no sperm competition tended almost universally to have smaller testes than the average. Harvey drew the conclusion that selection had produced this pattern, a conclusion made more confidently since in the same genera or families one would find different species on different sides of the line, divided by their different mating behaviors. The observed differences were too great to explain as a simple function of shared inherited traits (phylogenetic inertia) where selection was not involved. Since this initial study, many more analogous comparative studies have been done across a wide range of animal species, and the initial conclusions about testis size seem borne out more generally.

Optimality Models

Let us agree that often selection moves in diverse fashion. There is more than one way to skin a cat, and there are more ways than one to go from A to B. Horses run, humans walk, monkeys swing, kangaroos hop, birds fly, and snakes slither. But often (usually) we find that problem situations call for a more limited number of solutions, and perhaps just one is optimal. Natural selection is an incredibly powerful mechanism, and it does produce good solutions to design problems. Where a best solution exists, selection achieves it over and over again. All things being equal, selection does the same thing time and time again, to achieve the same ends.

The classic case is swimming. An organism needs a certain body shape with certain appendages to achieve the maximum effects—swimming quickly and moving up and down and sideways in the water. Not only has selection found this body shape but it has produced it repeatedly in many different kinds of animal—fish, reptiles, and mammals. The problems were the same, and the answers were the same. In the words of Richard Lewontin: "Adaptation is a real phenomenon. It is no accident that fish have fins, that seals and whales have flippers and flukes, that penguins have paddles and that even sea snakes have become laterally flattened. The problem of locomotion in an aquatic environment is a real problem that has been solved by many totally unrelated evolutionary lines in much the same way." Which leads to an obvious conclusion: "Therefore it must be feasible to make adaptive arguments about swimming appendages. And this in turn means that in nature the ceteris paribus [all things being equal] assumption must be workable" (Lewontin 1978, 228–229).

Let us exploit this fact as a general guide to all cases of adaptation, and not just those that have been repeated over and over in nature. Let us assume that selection has brought about perfect adaptation—it has "optimized" the situation—and that from here we can work out what is going on and why. Let us therefore build optimality models to explore cases of putative adaptation. The entomologists George F. Oster and Edward O. Wilson have likened their work on optimality models to that of an engineer. "In order to employ engineering optimization models the biologist tries to interpret living forms as in some sense the 'best.'" Of course, the trouble is with precisely what one means by "best" in a situation like this. "In effect the biologist 'plays God': he redesigns the biological system, including as many of the relevant quantities as possible and then checks to see if his own optimal design is close to that observed in nature."

It is all a matter of trial and error—of designing a model system and comparing it to the empirical findings. "If the two correspond, then nature can be regarded as reasonably well understood. If they fail to correspond to any degree (a frequent result), the biologist revises the model and tries again. Thus, optimization models are a method for organizing empirical evidence, making educated guesses as to how evolution might have proceeded, and suggesting avenues for further empirical research" (Oster and Wilson 1978, 294–295). Notice that Oster and Wilson fully acknowledge the heuristic virtues of optimality models. Using them spurs one on to further research and investigation. All is not simply intuition and bold confidence: the biologist-engineer attempts to go beyond mere story telling and to test his hypotheses. He runs up a model, derives predictions, and sees if they hold true. If so, all well and good, and if not, he tries a different model.

John Maynard Smith (1978b) has argued that these models do not test adaptationism as such. Rather, they test the models. "A particular model can be tested either by a direct test of its assumptions or by comparing its predictions with observation. The essential point is that in testing a model we are testing *not* the general proposition that nature optimizes, but the specific hypotheses about constraints, optimization criteria, and heredity" (Maynard Smith 1978b). But most would probably think it fairer to say that a two-way process is at work here. We adopt a background assumption of adaptationism and then we test specific models. Inasmuch as things work, then confidence in adaptationism builds or is confirmed. Inasmuch as things do not work, then we worry more about our adaptationist background as well as our model. It does not mean that we throw out the background as soon as we encounter problems. Darwinian adaptationism is our framework—what Thomas Kuhn would have called our paradigm—and without it we have no science to do.

Indeed, we feel justified in fiddling around with unsuccessful models to get them to fit the paradigm. This fiddling or adjusting does not take the work beyond the bounds of science. It is rather precisely what one does as a productive scientist. In listing all of the ways that adaptation could go wrong or fail, or fail to get measured, we are not simply acknowledging them and then forgetting them when it becomes convenient; biologists are not like Dickens's Mr. Micawber, who thought that listing his debts was as good as paying them. The exceptions and failures of adaptation happen for systematic reasons—comparative growth in the case of the Irish elk's antlers and lack of time to respond adaptively in the case of the dunnock's insensitivity to egg color. We assume adaptive perfection as a working hypothesis, we run up our models and test them, and if they fail then we use this failure as a tool to start exploring which factors on our list might be responsible for the failure. Ultimately, however, it is because things do work so well so much of the time that we can feel justified in the adaptationist paradigm and can so profitably seek out and find explanations for the exceptions. The strategy is methodological, not a blind metaphysical commitment to adaptation without exception.

Critics of adaptationism respond harshly to the use of optimality models. They complain that, for all of the talk of moderation, Darwinians acknowledge the problems with adaptationism and then promptly forget them. The philosopher Robert Brandon and the biologist Mark Rausher have this to say: "The attraction of optimality models is clear—they allow one to avoid history and genetics. Years ago in a discussion about number theory, Bertrand Russell said, 'The method of "postulating" what we want has many advantages; they are the same as the advantages of theft over honest toil. Let us leave them to others and proceed with our honest toil' . . . These are exactly our thoughts with respect to optimality models and the rigorous test of adaptationism" (Brandon and Rausher 1996, 200).

Can things really be this bad? Let us look more closely at a Darwinian using optimality models to try to understand his subject.

Fig-Wasp Sex Ratios

For our problem, we return to the question of sex ratios. We know that normally in sexual organisms one expects an even distribution between males and females. This is the evolutionarily stable strategy. As the ever-fertile mind of William Hamilton realized, lack of equal opportunity by males and females can skew this balance, however. Suppose, in particular, that one's sons are competing among themselves for the same females, and that no one else's sons are in the competition. Then really it would be a waste of your effort to produce many sons because from your perspective one son or the other is bound to reproduce, and it does not matter to you which one it will be—you are equally related to all of them. Hence, in such a situation, known as "local mate competition" (LMC), you expect a skew in the sex ratio toward more females. The experimental ingenuity of Edward Allen Herre and his associates allowed them to test this prediction.

Fig trees are widely ranged around the tropical parts of the world, and they have evolved in symbiotic relationship with various species of wasp, usually a particular species of wasp for a particular species of fig. The fig trees need the wasps to pollinate them. The flowers of the fig grow inside the fruit and, without some mechanical means of transfer, the pollen would not be carried from one tree to another. The wasps need the figs to protect and nurture their offspring. A fertilized female wasp will land on a developing fruit, enter it, and lay her eggs actually within the flowers. She then dies inside the fig, but the eggs develop and hatch—males first, which then go round inside the fig to find the females and fertilize them. The males soon expire, but not before cutting a hole in the wall of the fig, through which the females can escape and fly off to other fig trees. Although the wasps are very small, only a few millimeters long, and have life spans of only a few days, they can make use of air currents and fly large distances (twenty kilometers or more). If more than one female invades the same fruit (which they often do), they are highly unlikely to be related.

One can make a number of predictions about sex ratios in this situation, because the males are all confined within the same space and are competing for the same females. Three forecasts are highlighted by Herre. First, following directly from Hamilton's insight, there should be a female bias in the offspring being produced. Second, assuming that the females can control the bias, the degree of bias should increase as the number of foundresses decreases. If just one foundress has entered the fig, then the males are all brothers and very few of them are needed. If there are several foundresses, however, then the sons of different females are competing and more males are needed to enter the fray. A third prediction is connected with the fact that the wasps are hymenoptera, which means that females have two parents whereas males have only mothers. If inbreeding occurs, in which females mate with their brothers, then this will increase the relatedness of mothers to their daughters (but not to their sons), and for this reason one expects a bias toward female offspring, who carry more copies of the mother's own genes.

The fig/wasp connection makes an ideal subject to test these claims. The female foundresses die and leave their corpses within the fruits, so the number of females that have laid eggs is readily known. The offspring can be sexed and counted. And increasingly one can do molecular analyses to check relationships. All of which leads to most impressive results: the wasps do indeed show sex ratios biased toward females; they do show less bias when the number of foundresses go up, thereby increasing the competition between unrelated males; and they do show sex biases as the result of inbreeding.

Other predictions are also possible. Where there is one foundress only, ideally she would have only one son; but she cannot risk having no one to pierce the fruit and allow the females to escape. An insurance factor in the form of a second male must be built in, although as the number of offspring grows this factor will be proportionately less. (Two males in a group of ten counts for more than nineteen males in a group of one hundred.) This factor can also be found and mapped.

A triumph of optimality theorizing! Well, not exactly. According to Herre and colleagues, "Although there is qualitative fit to all of the theory's basic predictions, there is considerable variability among species in the sex-ratio responses associated with different numbers of foundresses. Some species show pronounced shifts in sex ratio and others show almost none. Moreover, the sex ratios for multiple-foundress broods generally show more female-biased sex ratios than the unadjusted model predicts" (Herre et al. 1997, 233–234). This leads Herre into considerable philosophical angst of the kind we have already seen consuming Maynard Smith. Which is to come first? The theory or the evidence? Do we tinker with the theory or do we excuse the findings? Is adaptation our governing rule, or is adaptation itself under test? Are we now to throw out our theory and seek other answers for our results? Or are we to keep our theory and explain away the results as anomalies?

In fact, a bit of both goes on, as is usually the case. The failure of fit seems to come most often in those cases where things are not commonplace. Herre ran a number of experiments where female wasps were forced into founding colonies in figs that they never use in nature. This led to all sorts of anomalies between fact and theory, confirming an adaptive lag between the actuality and the ideal, during which organisms are moving to the optimum through selection but often are still in a stage of transition. The researchers point out that they are not working in complete isolation in making such suggestions. They can legitimately draw on the cumulative knowledge of evolutionists about such situations. "This interpretation is appealing because it is consistent with a wide range of studies that demonstrate that organisms are physiologically best adapted to the environments they encounter most frequently, and develop the most plasticity in the most variable environments" (Herre et al. 1997, 234).

Still other questions have been asked and answered. Could it be that the wasps are simply showing the effects of adaptations from the past and not the present? We know from experimental evidence that sex ratios are heritable and that they respond rapidly to selection. Using molecular data, the relatedness of the wasps could be compared and a phylogeny constructed. "The simple and clear answer was that there is no evidence at all that wasps of one kind (say, given to single foundress situations) were all of one shared ancestry and wasps of other kinds had other shared ancestries. We seem rather to have cases of independent evolution, according to individual needs and problem situations."

All of which starts to point Herre and his colleagues to this conclusion: "If we define 'adaptationism' as 'the proposition that organisms are perfectly adapted in all respects for all situations that they encounter' (compare Orzack and Sober 1994 a, b), then, in our opinion, 'adaptationism' is a ludicrous proposition and not worth testing. To believe in the strong form of this proposition is to ignore the ubiquitous variation in morphology, physiology, behavior, genetics, survival, and reproduction that is characteristic of any natural population." They continue: "It is the variation in these attributes and the form of the relationships among them that make evolution by natural selection possible. If we dismiss adaptationism thus defined, the important question shifts from 'Optimality: yes or no?' to 'What situations favor relatively greater precision in adaptations?' As we have shown with the examples discussed here, optimality models can be productively employed toward the end of addressing the form and even the precision of adaptation" (Herre, Machado, and West 2001, 214).

Coda: "The attraction of optimality models is clear—they allow one to avoid history and genetics . . . 'The method of "postulating" what we want has many advantages; they are the same as the advantages of theft over honest toil. Let us leave them to others and proceed with our honest toil' . . . These are exactly our thoughts with respect to optimality models and the rigorous test of adaptationism." No comment.

CHAPTER 11

FORMALISM REDUX

CHARLES DARWIN'S EVOLUTIONISM insisted on the primacy of function. Let us now turn to the other side and see what if anything remains of the formalist tradition. If we grant the great significance of function, what then of form?

Morphology—which compares anatomical characteristics in order to discern relationships—is bound to emphasize form over function. Traditional morphologists such as Geoffroy Saint-Hilaire, Richard Owen, and Thomas Henry Huxley were looking for stable features, conserved over the long term, that reflect old relationships, and they tried diligently to avoid or discount the ephemeral—characteristics that are recent and flexible. After Darwin, the practices of the morphologist were reinterpreted in an evolutionary context—the older the feature, the longer since the break between organisms now very different. But the very fact of evolution makes the morphologist's day-to-day task more difficult. Methodologically, the morphologist is trying to downplay change. The more unchanging the type, the easier it is to discern which organism is connected to which. For this reason, although at one level the arrival of evolution was profoundly revolutionary for the study of form, at another level it made virtually no difference at all.

Nearly a hundred years ago, Edward Stuart Russell, who is still the best historian of the form/function relationship, wrote: "We shall see that the coming of evolution made surprisingly little difference to morphology, that the same methods were consciously or unconsciously followed, the same mental attitudes taken up, after as before the publication of the *Origin of Species*" (1916, 247). The fact that hardly anyone accepted Darwin's mechanism, natural selection, as the significant cause of evolutionary change just reinforced this attitude. If function was irrelevant, then so also were mechanisms that acted on function. Even when people thought about change, they downplayed adaptation. Although *Naturphilosophie* was developmental through and through, it considered the type rather than the end as the fundamental organizing concept.

Little wonder, then, that when the synthetic theory of Darwinism appeared on the scene in the 1930s and 1940s, morphology was pushed aside and ignored, except in certain antediluvian circles on the Continent. As the morphologist Michael Ghiselin (1980, 181) noted rather sadly about his own trade: "Morphology has contributed so little primarily because it has had so little to contribute. It is a descriptive science of form, and only when conjoined with other disciplines does it tell us anything about causes. But once a causal mechanism has been accepted, it can provide a valuable service. Nonetheless, for this very reason, morphology tends to be the sort of discipline that will follow, rather than lead, in the development of evolutionary theory."

Now, however, as we move into the twenty-first century, a dramatic sea change has occurred in the fortunes of morphology. The study of "type" is making a major comeback, and many biologists are pushing form over function. Various reasons account for this new zest for structuralism, but the buzzword today is "constraint"—that which supposedly pushes selection to one side, as unbreachable laws of nature unfold and shape the organism.

What is a constraint? For all of his Darwinian commitments, John Maynard Smith (Maynard Smith et al. 1985, 269) has well explained what is at issue: "Organisms are capable of an enormous range of adaptive responses to environmental challenge. One factor influencing the pathway actually taken is the relative ease of achieving the available alternatives. By biasing the likelihood of entering onto one pathway rather than another, a developmental constraint can affect the evolutionary outcome even when it does not strictly preclude an alternative outcome." One of Maynard Smith's co-authors, the Spanish morphologist Per Alberch, illustrated the notion of constraint and its effects by positing a population of organisms that can be broken down into several distinct morphological groups. Why the different groups and in particular why the gaps between the groups? Go back in time to a supposed unbroken, seed population. The selectionist supposes that the subsequent populations and the gaps that separate them reflect adaptive highs and adaptive lows: empty spaces represent organisms that could not survive and reproduce with the characteristics represented by the gaps.

To the contrary, the constraint enthusiast supposes that the present actual distributions exist not because the kinds of organisms represented by the gaps are maladapted in some way but rather because they could never become those organisms in the first place. If organisms could ever get to the places where there are gaps, they would be just as fit as everyone else; the point is that they cannot get there. Suppose that the seed population were allowed to grow and diversify without selection having any shaping effects. For the selectionist, the subsequent population would be one large unbroken group, spread all over—the adaptive landscape would be a plain, with no highs or lows. For the constraint supporter, the absence of selective forces would make no difference; the present groupings would still have evolved.

Let us turn now to look at some of these constraints and the ways in which function is supposedly driven back and confined by form.

Genetic Constraints

Whether or not it is the most fundamental, the lowest level of constraint operates at the genetic level: something originating at the physical or chemical level of the gene affects the evolutionary path taken by the gene's possessor. Ultimately, if the genes cannot produce a certain product, then the organism cannot use that product to its adaptive advantage. So, as we have seen, if a mutation producing a variation for a particular color, for example, does not occur, then that color will not be available to the organism. Also, if the genes can produce one product only if they also produce a second product simultaneously (what we call pleiotropy), then the organism must weigh the advantages of the one product against the disadvantages of the other.

However, as we have already seen, how much of a constraint these and related factors really are in nature is not so clear-cut. Often, if an organism cannot get a product one way, it can get it in another way, or it can find a substitute that might already be held in the population by one or another kind of balancing selection. And in the case of pleiotropy, where the same gene produces two effects, one presumes that normally the effects come because the gene produces (or fails to produce) some cellular substance—a protein. If one of the effects really is deleterious but the other is highly advantageous, under strong selective pressure the organism will often enlist other genes to modify the active gene producing the product or block the troublesome product and hence eliminate its unwanted side effect.

The developmental morphologist Rudolf Raff (1996, 304) raises another genetic constraint: genome size. "Having a large genome has consequences outside of the properties of the genome per se." Having a large genome size—that is, having lots of DNA—in itself puts constraints on an organism. For starters, having lots of DNA means that an organism is going to be slower replicating its genome than if the DNA were less. Hence, overall growth might well be slowed. In addition, large genomes mean large cells, which in turn mean that the cell-surface-to-volume ratio is decreased (surface increases as the square whereas volume increases as the cube), and this can also slow down metabolism.

Salamanders often have large genomes and thus are good organisms on which to test hypotheses about this constraint. And some evidence of their operation has been found. Having said this, however, Raff has to admit that if such constraints are at work, they apparently do not make much difference. The salamanders can do some pretty remarkable things—remarkable salamander things, that is—and seem to be not at all functionally constrained: "These salamanders occupy a variety of cavernicolous, aquatic, terrestrial, and arboreal habitats. They possess a full range of sense organs, and most remarkably, a spectacular insect-catching mechanism consisting of a projectile tongue that can reach out in ten milliseconds to half the animal's trunk length (snout to vent is the way herpetologists express it)." They have pretty good depth perception, too. And indeed, the slow metabolic rate brought on by their large genome size may even be of adaptive advantage. "The low metabolic rates introduced by large cell volume may be advantageous to sit-patiently-and-wait hunters that can afford long fasts. Vision at a distance is reduced to two handbreadths, but since these animals are ambush hunters that strike at short range, that probably doesn't affect their efficiency much" (p. 306). All in all, an ardent selectionist does not find much to worry about here. Apparently, if need be, salamanders can even start to reduce their genome size. The genetic constraints are just not that strong.

Historical Constraints

One of the most discussed notions put forth by the new formalists is that of historical or phylogenetic constraints. This is the idea that some feature of an organism—a feature that first appeared way back—is now ingrained in the organism's development at the most fundamental levels and is therefore impossible to shake in subsequent generations. (This is sometimes spoken of as a developmental constraint, in reference to the fact that it operates during the organism's development; generally, however, a developmental constraint is a somewhat broader notion and covers other kinds of constraint, to be discussed shortly.)

Historical constraints take us right back to the traditional issues about form and function, because at some level today supposedly we are presented with homologies based on features of the past—homologies that now have little or no adaptive function and perhaps even steer organisms away from their adaptive peaks. The vertebrate skeleton is still a favorite example. Could it be that the four-limbedness of the vertebrates—their *Bauplan,* to use the fashionable term that has replaced archetype—is a less-than-optimally-adaptive legacy that we all have to live with, even though we would have been better off with six limbs, like angels and insects?

This discussion takes advantage of exciting discoveries in molecular embryology. At the genetic level, organisms seem to be even greater recyclers of already available material than they are at the level of organs or bones. Most remarkable of all are certain "homeotic genes." These are not structural genes—that is, genes coded to make actual body proteins—but developmental genes coded to process the products of structural genes. The homeotic genes regulate the integrity and order of the parts of the body—a mutation in one perhaps moving an eye to where a leg might normally appear or vice versa.

A subclass of homeotic genes consists of the *Hox* genes, found in bilaterans (organisms that are the same on both sides). *Hox* genes order the appearance of various body parts and seem to work in the same sequence as they are found on the chromosomes. In *Drosophila* (fruit flies), the *Hox* genes start at the head, work down through the thorax, and so on to the end of the abdomen. Within these genes, one finds lengths of as much as 180 base pairs that are used to bind the genes to other DNA segments that are part of structural genes. In other words, these "homeoboxes" make a protein (of sixty amino acids)—the "homeodomain"—that is the key to the *Hox* genes' ability to regulate the structural genes.

Absolutely staggering was the discovery of a homology between the homeodomains of fruit flies, frogs, fish, mice, and humans. Although it has been many hundreds of millions of years since humans and fruit flies shared a common ancestor, we nevertheless use essentially the same chemical mechanisms to order the production of our various body parts. The fruit flies' legs and the humans' legs are produced through very similar processes. The similarities are just too great to be the result of chance or of convergence around a common (best) solution. These similarities are homologous, and they have led some biologists to downplay entirely the significance of selection. "The homologies of process within morphogenetic fields provide some of the best evidence for evolution—just as skeletal and organ homologies did earlier. Thus, the evidence for evolution is better than ever. The role of natural selection in evolution, however, is seen to play less an important role. It is merely a filter for unsuccessful morphologies generated by development. Population genetics is destined to change if it is not to become as irrelevant to evolution as Newtonian mechanics is to contemporary physics" (Gilbert, Opitz, and Raff 1996, 368).

The science behind these assertions is terrific—no question about that. From the genesis of the synthetic theory of Darwinism around 1930, the one great missing element has been developmental biology or embryology. Now it is assuming its rightful place, just as Darwin himself would have wanted. But is developmental biology pushing selection to one side, as its proponents argue? Are we being thrown back to the post-Darwinian period, when some evolutionists thought embryology was the key to change and considered selection to be a minor and irrelevant factor? Or is an enthusiasm for the newest biology on the block overtaking a more balanced and just assessment of what it all means?

Let us go back to the homologies of the vertebrate skeleton for a moment. No one will deny that they exist and are obviously important—a clear sign of evolution and a significant aspect of organic form. But do they, or any similar instances of phylogenetic inertia, constrain evolution in any significant way? Or are they instances of an optimal solution? By analogy, could it not be that "the fact that all tires are round more likely means that round wheels are optimally functional than that tire companies are somehow constrained by the round shape of their existing molds. Thus phylogenetic inertia is not an alternative to natural selection as a mechanism of persistence, and evidence of the former is not evidence against the latter" (Reeve and Sherman 1993, 18). One needs evidence of more than homology to argue against adaptation. One needs evidence that homology is *preventing* adaptation.

Darwinians suspect that sleight of hand is at work here. The new formalists, people like the late Stephen Jay Gould, focus on things like the four-limbedness of vertebrates, which have no obvious adaptive function today. Rather than plunging into a quest for adaptation in the past, when such features first appeared, the formalist focuses on the present and—thanks to somewhat idiosyncratic (and self-serving) definitions of adaptation that refuse to apply the term to subsequent uses after the initial use—all four-limbedness today is labeled as nonadaptive. And from there it is all too easy to slip into calling it "maladaptive." A final move is to appropriate some fancy name like *Bauplan,* to give ontological status to what you are promoting, and the dish is complete. Function is relegated to the sidelines.

In George Williams's (1992) opinion, the notion of a *Bauplan* is not only "misguided and dispensable" but cuts off further debate and inquiry. What right has one to say that homologies, or *Baupläne,* are so very nonadaptive? Certainly, even though the formalists refuse to look into these matters, the rest of us can inquire whether the original key ancestors were using the relevant features in a nonadaptive fashion.

This kind of inquiry dredges up much worthwhile information. As Maynard Smith has pointed out, "The basic vertebrate pattern arose in the first place as an adaptation for sinusoidal swimming. Early fish have two pairs of fins for the same reason that most early aeroplanes had wings and tailplane: two pairs of fins is the smallest number that can produce an upward or downward force through any point in the body."

More than this, although homologies as such may not today have an adaptive function, one should not assume that they are so very constraining, certainly not constraining in a way that moves organisms away from adaptation: "Some of the earliest vertebrates had more than two pairs of fins (just as some early aeroplanes had a noseplane as well as a tailplane). Hence there is no general law forbidding such organisms" (Maynard Smith 1981, 11). Even that classic constraint of constraints, the vertebrate hand or paw, seems to have more evolutionary flexibility than one might suppose. We all know that the five digits can be reduced in number—the horse proves that. But what about going up in number? In fact, one fairly often does get people who have mutated up a digit—Robert Chambers was one such person. And it now appears that some early land vertebrates had seven or eight digits. Not much channeling here.

Does the fossil record give any support to historical constraints? Some have certainly thought that it does. Gould and Eldredge became famous for their claim that the fossil record does not exhibit smooth uninterrupted change but shows organisms in extended periods of uniformity (stasis) followed by abrupt changes from one form to another. It shows a pattern of "punctuated equilibrium," which could be evidence of non-Darwinian processes of change. Whatever the causes of the abrupt change, the long-term uniformity comes because organisms are locked into place by their historical forms, they argue, and only the most violent shaking can shift organisms from their nonadaptive stability.

Most orthodox Darwinians grant that the fossil record often, if not usually, exhibits the kind of stasis that Gould and Eldredge claimed, although certainly there are many exceptions. Still, a fairly traditional Darwinian explanation is at hand. Apart from the fact that a great deal of change is possible that would not be reflected in the fossil record—skin color, for instance, as well as quite significant fluctuations over micro-periods of mere thousands of years—a well-known form of selection works to stabilize by favoring the average or the mean. A notable case is birth weight in humans. Those infants with average birth weight have a better chance of surviving than those who are either much larger or much smaller. This reversion to the mean accounts for the stasis in large populations, and the founder principle or something similar accounts for the rapid change that occurs in smaller populations. Normally change is not needed, and selection resists it. But when change does occur, selection brings it about rapidly, and afterward organisms settle down again into a stable, successful pattern.

Of course, the orthodox Darwinian can still claim that this is all a storm in a teacup. The fossil record really is very fragmentary and the traditional explanation of the staccato pattern—it is an artifact of lost data—is the right one. Or, as Douglas Futuyma has argued, it may simply be that normally any changes in a species get wiped out over time, not merely because of fluctuating circumstances (and hence fluctuating selection pressures) but because of the fact that the reproduction occurring within a species evens out all the short-term changes. It is only when for various reasons, such as travel by a subgroup to an isolated area like an island, a species gets broken and reproductive isolation occurs that (even without the founder principle), thanks to selection, changes now start to accumulate within the separate populations, and this then gets to make its mark in the fossil record. "Although speciation does not accelerate evolution within populations, it provides morphological changes with enough permanence to be registered in the fossil record" (Futuyama 1987, 467).

In Gould's last major work on evolution and its causes—*The Structure of Evolutionary Theory*—Gould endorsed this way of thinking. Given that it is as traditional a Darwinian, selection-dependent answer as one could imagine, one is left not knowing whether to admire Gould's chutzpah at making Darwinism the linchpin of his speculations or to mourn the passing of his non-Darwinian challenge. In any case, to summarize, punctuated equilibrium may point to some significant nonselection-driven causal factors. It may point to some significant selection-driven causal factors. Either way, constraints are not among them.

Finally, what of the new molecular discoveries? Do they change the picture at all? Raff, who has spoken eloquently about the triumph of the developmental way, nevertheless has pointed out that genetic homologies may in fact not be all that rigid. He discusses a certain *Hox* gene (the Antennapedia complex) that occurs in two species of fruit flies, *D. melanogaster* and *D. pseudoobscura.* As measured by molecular clocks, these two species parted company about 46 million years ago—not that far back in the history of life. As you might expect, the Antennapedia complex in the two species is remarkably similar, far too similar to be there by chance. But the complex in the two is not identical. In *D. pseudoobscura,* one gene, the so-called Deformed gene, is in the same place and has the same orientation as that of mammals. In *D. melanogaster,* however, its orientation is reversed. Turned right around. Other similar differences have been found, adding up to Raff's conclusion that "the Hox complex can tolerate substantial change over moderate periods of evolutionary time and within a common body plan" (1996, 309).

As I have said, the science is terrific. The insights into evolutionary process are magnificent—unexpected and revealing. Whether they spell the demise of Darwinism with its focus on function is another matter altogether.

Structural Constraints

Let us look next at what have been labeled structural constraints, although these too would seem to fall under the heading of developmental constraints. Here, the very task of putting together a functioning organism is the constraining factor. Once again Gould, with Richard Lewontin, raised the famous question. In a celebrated critique in 1979, Gould and Lewontin drew attention to the triangular spaces at the tops of the four pillars that hold up the circular dome in medieval churches. These spaces are usually highly decorated, often with brilliant mosaics of the evangelists (as at the Cathedral of San Marco in Venice, which inspired Gould's analogy). Gould and Lewontin argued that although such spaces ("spandrels") seem to have been designed as a space for displaying the creative outpourings of artisans, in fact these spaces are just by-products of the builders' design for holding the roof up. Once a dome held up by four rounded arches with pillars was chosen by the builder, the triangular spandrels appeared as a necessary consequence in the four corners, above the pillars. But once they did appear, artisans took up the task of integrating these spaces into the cathedral's decorative scheme by filling them with mosaics. "The design is so elaborate, harmonious, and purposeful that we are tempted to view it as the starting point of any analysis, as the cause in some sense of the surrounding architecture" (p. 148). This, however, is to put the cart before the horse. And perhaps, argue the authors, we have a similar situation in the living world. Much that we think of as being adaptive is merely a flashy spandrel produced by constraints on development and not an example of optimal design. Perhaps things are much more random and haphazard—showy but ultimately nonfunctional—than the Darwinian thinks possible.

To which comes the response: did any Darwinian ever think otherwise? After all, allometry has long been cited as a prime example of the difficulty of putting a functioning organism together properly and optimizing every last feature. The Irish elk had a terrific early sex drive and a lousy later hairpiece, or so many think. Darwinians have always agreed that organisms tend to have some things left over, and that these redundant or unwonted characteristics might later be picked up by selection and used in their own right—as of course the spandrels of San Marco were. Pumping testosterone through the bodies of male humans has all sorts of adaptive advantages—penises and testicles for a start. Perhaps hair on the male face is but a secondary effect of that testosterone surge, of no adaptive advantage; perhaps, but despite the best advertising efforts of the Gillette razor company, this has not yet been proven to be so. Or to use Gould's own favorite example, the much enlarged clitoris of the female hyena, which he claimed had no present adaptive significance: the Darwinian would agree that this pseudo-penis may have come about by chance, but this does not mean that it has no value today in mating rituals. In fact, a central tenet of Darwinism holds that unexpected side effects can be, in the long run, a crucial part of the evolutionary story.

The really big question is about how new characteristics ever get started in the first place. As Gould forever asked, could a tenth of an eye be of any great value? Well, as Dawkins responded, perhaps it could, but there is no need to suppose immediate value in every new feature. Feathers today obviously have the adaptive edge when it comes to flying, and no one would say that flight is without its purposes. Increasingly, however, the evidence suggests that feathers first appeared on the dinosaurs not for flight but for other ends, most likely insulation and heat control. Only later were they used to invade the air.

But could it not be that Darwinians are missing the forest for the trees? Could cultural phylogenetic inertia be blinding them to structural constraints that can be very important? No one denies that such constraints lead to new instances of adaptive excellence. The point is whether they introduce a whole new dimension into the discussion, by showing that much in the organic world is fundamentally nonadaptive. Darwinians have failed to see this and still continue not to see it. For Gould, one subject especially interesting in this respect was *Homo sapiens.* He argued that, from a biological perspective, human nature makes sense only if we recognize that much of our thinking apparatus and its consequent production of culture is in fact a spandrel—that it emerged from the evolutionary process but was not under the control of natural selection: "The human brain may have reached its current size by ordinary adaptive processes keyed to specific benefits of more complex mentalities for our hunter-gather ancestors on African savannahs. But the implicit spandrels in an organ of such complexity must exceed the overt functional reasons for its origin . . . I suspect that many puzzling features of human mentality would be better resolved if we conceptualized them as historical constraints derived from distant adaptational origins." Gould concluded: "Any 'evolutionary psychology' that neglects the nonadaptational origin of many features now useful (or at least used, however dubiously), and that limits the domain of evolutionary inquiry to arguments (often speculative) about initial adaptive causes and benefits, will become more misleading than enlightening in restricting investigation to such a narrow scope of inquiry. We must abandon the largely unconscious bias of an overly strict Darwinian approach that equates all 'evolutionary' explanation with adaptationist analysis" (Gould 2002, 1264–1266).

Could it be true that human intelligence and the consequent culture is a nonadaptive consequence of other biological pressures, and perhaps reflects this today? It is certainly hard to imagine that in the past four million years of human evolution—from Lucy, with a brain of less than 500 cc, to humans, with a brain of about 1200 cc—no adaptive forces have been at work on the human mind. The most recent thinking of anthropologists is that our ancestors faced a highly variable climate and would have found great advantage in having the mental flexibility to respond to these changes. Brain also helps if one has to search for food and provide defense against attack—especially against other intelligent animals after food and mates, namely other hominids. Yet for all that selection may have been the main factor in the growth of human brainpower, no one—not even the most ardent Darwinian—is going to argue that every last bit of culture today is adaptive.

Consider the different ways in which the English and the French spoken languages convey plurals. Compare "the boys" with "les garçons." For the English, the pronoun stays constant from singular to plural and the needed information comes at the end of the noun, with a buzz sound for the plural. For the French, singular and plural nouns sound alike (though spelled differently) and the information is conveyed in the different pronouns, "le" and "les" (pronounced *luh* and *lay*). Even if a form of cultural selection was at work in producing these kinds of differences, natural selection is most surely not involved.

On the other hand, few would deny biology some significant role in culture, or at least in human behavior and thought. Whether you choose a sonnet or blank verse to express your love may not be influenced by your biology, but the emotions you feel are pure Darwinism. The question is really about the area in between and the extent to which it is adaptive and the extent to which it is not. Darwinians who work on these issues, the evolutionary psychologists, generally take a middle position, as did those workers who study infanticide. Almost no evolutionary psychologist would say that the surface of culture—what one might call the flesh and skin—is tied directly to adaptive advantage, but all would argue that the underlying rules and patterns of behavior—the skeleton—is rooted deeply in adaptive advantage. One way of expressing this is by arguing that we have certain innate dispositions (what have been called "epigenetic rules") that are rooted in biology; but the actual manifestation of these dispositions takes us into the realm of the nonbiological, dependent on the situation, background, and other influences.

Consider moral prescriptions and behaviors, which are a fundamental part of culture: "Killing is wrong." "Love and cherish your children." "Give to the poor." "Do not have sex with someone else's spouse." Some people argue that morality has little or nothing to do with the genes. They point to the relativism of different societies (in parts of Africa, female circumcision is a good thing and in all of North America it is not) and to the nonadaptive nature of much morality (the soldier who obeys duty and goes over the top of the trench in the Battle of the Somme) and they conclude that morality and biology have little or nothing to do with each other. This is the position of many social scientists, as well as Marxists like Lewontin.

Others see morality as a nonadaptive spandrel—at least a nonbiologically connected spandrel—that is nevertheless rooted in features that evolved for other adaptive reasons. This seems to be the position of the geneticist Francisco Ayala (1987, 239): "Ethical behavior is an attribute of the biological make-up of humans and, hence, is a product of biological evolution. But I see no evidence that ethical behavior developed because it was adaptive in itself. I find it hard to see how *evaluating* certain actions as either good or evil (not just choosing some actions rather than others, or evaluating them with respect to their practical consequences) would promote the reproductive fitness of the evaluators."

Finally, the hardline Darwinians argue that ethics is directly connected with biological need. Edward O. Wilson and I have argued just this. Like Darwin, and for much the same reasons, we think that being ethical has a direct biological payoff. We find ourselves unmoved by the relativism of ethics. In the past fifty years, the West has seen major shifts in the direction of sexual liberation and freedom, for instance with respect to toleration and acceptance of homosexuality. But this is less because of any fundamental change in morality at the base level—certainly not because of any biological change in human nature—and more because we now have greater understanding of homosexuality. It is now appreciated that sexual orientation is not usually freely chosen (and hence should not be considered sinful), nor is it necessarily the result of dysfunctional families (and hence should not be considered an illness and a subject for medical intervention and avoidance). And the supposed counter-examples fail to persuade. Most of our lives we are not going over the top in the Battle of the Somme, and when people did, it was for very human (that is, plausibly biological) reasons. Not wanting to let down your buddies seemed to be the main motivation, though knowing that you would be shot by your own commanding officer if you refused might have weighed in the decision. In any case, no one ever said that adaptations are perfect. Just better than the alternative. Moral rules function rather like the breeding rules for Davies's dunnocks. Generally they work, but nature can be fooled.

For Lewontin, therefore, talk of innate dispositions is unneeded and confusing. For Ayala, perhaps such dispositions exist, but they seem today to have little to do with direct adaptive advantage. For Wilson and myself, such dispositions underlie morality but express themselves in different (although constrained) ways in different situations and cultures. Let us leave it at that—although frankly, whether the language of constraints and spandrels throws much new light on these issues might be doubted.

Physical Restraints

Here we come to something that no one will deny. Here constraints really do matter. The physical world influences organisms and sets limits on what organisms can and cannot do, and where and when they should invest their energies. Take one of the most basic of all facts, that size and consequent weight go up rapidly according to the cube power of length or height. Suppose you have two identically shaped mammals, one twice the height of the other. It is going to be eight times as heavy. This means that, from a structural perspective, it has got eight times the weight problem. You simply cannot build elephants as agile as cats. Because of the effect of gravity, elephants need far more support to stay upright, which in turn means bigger and heavier bones.

Physical restraints lead to some interesting consequences for the organic world, showing how very different this world is in many respects from the human world. Organized complexity in the rest of the organic world is not the same as organized complexity in humans. For instance, it is much easier for nature to make a circle than a square, a sphere than a cube. Circles and spheres minimize the area-to-volume ratio and make for even forces all around (think of soap bubbles). This means that, while humans are obsessed with right angles, in nature you rarely find right angles. They are not impossible, however: pine trees stand up in the forest perpendicular to the horizontal floor, dividing eggs cleave on lines at right angles, and the inner ear has canals at right angles to detect acceleration. Most fascinating and unsuspected are square bacteria from saline pools in the Sinai Desert. They lie flat on the surface of the water, in sheets of up to eight or sixteen, rather like postage stamps. Because of the salinity of the water, the bacteria have no excess internal pressure that might make them bulge into circles or spheres. Apparently, there is no positive or negative adaptive value to their shape.

The interesting question is not so much why nature contains so few right angles but why humans are so right-angle obsessed. Probably it is the conceptual ease that such thinking conveys. Imagine trying to design houses where the key unit of measure was 95 degrees, or where the pentagon replaced the square. Although, even here, right angles have their costs. Think of how much cross-bracing is needed to get things to hold in shape. Four strips of wood nailed together in a square is far less rigid than three pieces of wood nailed together in a triangle.

Apparently then, even physical constraints allow dimensions of freedom that might not be intuitively obvious or that one might think, a priori, not to be possible. More than this. As in the case of other kinds of supposed constraints, sometimes it is not obvious that physical constraints should really be called "constraints." John Maynard Smith and his colleagues have explored in depth the example of the coiling of shells in molluscs and brachiopods. The coiling itself is fairly readily reduced to a simple logarithmic equation, and it is possible to draw a plane that maps the coiling as a function of the vital causal factors, particularly the rate of coiling and the size of the generating curve. On such a map, one feature stands right out for comment: whereas for most shells the coils touch all the way from the center to the perimeter, some such shells coil without touching. There is a gap between the coils.

In real life you would expect that organisms would fall on the side of the map covering shells that touch as they coil. Touching gives greater strength and at the same time conserves material, as the outer surface of the inner whorl functions as the inner surface of the next outer whorl. When Maynard Smith and colleagues mapped the actual shells of a group of organisms, the genera of extinct ammonoids (cephalopod molluscs), the fit between the theoretical and the actual was outstanding. The ammonites almost all fell on the side of coil touching. Even the exceptions proved the point. The shell of the living pelagic cephalopod

Spirula has a shell that coils but does not touch. But this organism carries the shell internally, using it for buoyancy. There is no need for strength.

Maynard Smith and colleagues concluded "that the constraint against open coiling is an adaptive one brought about by simple directional selection." A conclusion which surely brings us full circle, for if constraints can be adaptive, brought on by selection, the distinction between form and function has truly collapsed. The theoretical biologist Gunter Wagner goes so far as to argue that constraints may even be necessary for the action of selection, else variation will be all over the place, with any positive changes being outweighed by other moves in a maladaptive direction.

Order for Free

We have gone from physical restraints being opposed to selection to physical restraints working hand in hand with selection. Could it be that physical restraints entirely replace or significantly supplement selection? Whether or not one is now directly thinking in terms of adaptation, or at least whether one is thinking in terms of sophisticated adaptation, could it be that physics alone does the job of organic design? Right through the twentieth century some people thought that this was so. The most prominent was the Scottish morphologist D'Arcy Wentworth Thompson, whose *On Growth and Form,* first published in 1916, was a classic defense of form over function, worthy of being on the shelf with anything written by Goethe, Geoffroy, or Richard Owen. Although an Aristotelean, Thompson had little sympathy with the Darwinians' focus on final cause and argued strenuously that (what Aristotle would have called) material and formal causes are prior and more important. What really counts is the physics of the materials being used to make organisms and the mathematical laws that govern these materials. You build a house because you want to live in it, but the building of the house is governed by physics and mathematics. The same is true of organisms.

Exactly how evolution was supposed to take place was somewhat fuzzy in Thompson's world. He was not against adaptation as such, but he did relegate it to being an obvious corollary to the development of form as governed by the principles of physics and chemistry. A beautiful example of his kind of approach is given by his analysis of the shapes of jellyfish. For him, these came about simply through the physics of drops of liquid of one density falling in a liquid of a different, somewhat lower density. As the patterns of the falling liquid, so the patterns of the organic, liquid-dwelling jellyfish. For Thompson, the analogy clearly implied that, whatever the force driving evolution, it was constrained by the laws of physics and chemistry. But this is no worry, for these laws not only constrain but build and create. The adaptations of the Darwinian—at least the only adaptations that the biologist need take seriously (structure-based adaptations like wings and limbs, as opposed to what he regarded as trivial surface adaptations like coat color)—emerge effortlessly from the way that the world of physics and chemistry is destined always to run.

In recent years, starting with Gould, who wrote an early appreciative essay, Thompson's approach has found enthusiastic supporters. In England, the Canadian-born morphologist Brian Goodwin has been prominent among them, as has been the medically trained theoretical biologist Stuart Kauffman in the United States. Kauffman has been the path-maker into the thickets of a concept known as self-organization or, as Kauffman cleverly calls it, "order for free." "The tapestry of life is richer than we have imagined. It is a tapestry with threads of accidental gold, mined quixotically by the random whimsy of quantum events acting on bits of nucleotides and crafted by selection sifting. But the tapestry has an overall design, an architecture, a woven cadence and rhythm that reflect underlying law—principles of self organization" (Kauffman 1995, 185). A simple example of such self-organization, accepted by all, is yielded by the so-called Beloussov-Zhabotinsky reaction, a phenomenon discovered in Moscow in the 1950s. When a mixture of organic and inorganic substances is placed on a flat plane (as in a Petri dish), they make concentric rings that move out from the center and vanish as they encounter other such rings. On its own and in itself, this is little more than a pretty pattern. But it so happens that these kinds of rings are seen also in nature. In particular, the cellular slime mold goes through a phase in which it simulates the Beloussov-Zhabotinsky reaction very precisely. Such slime molds are usually a colony of free-living amoebas, eating bacteria. But if food supplies become scarce, they begin to aggregate. "Cells start to signal to one another by means of a chemical that they release. This initiates a process of aggregation: the amoebas begin to move toward a center, defined by a cell that periodically gives off a burst of the chemical that diffuses away from the source and stimulates neighboring cells in two ways: (1) cells receiving the signal themselves release a burst of the same chemical; and (2) they move toward the origin of the signal" (Goodwin 2001, 46). What is remarkable is that, as these amoebas begin to move together—and finally combine into a multicellular organism that can fruit and reproduce, making another crop of independent amoebas—the patterns of movement they exhibit are identical to those of the Beloussov-Zhabotinsky reaction.

In fact, the molecules in the chemical state and the living state are quite different, but the underlying feedback process is similar, where substances are produced in increasing amounts until other processes take over to inhibit the production of these substances. Then, the whole system exists in an unstable condition of oscillation, as the various processes switch on and off. For the new formalists, we clearly have a case where an organism (or group of organisms, depending on how you count the slime molds) uses a self-generating, chemical process for its own biological ends. The overt pattern was not shaped by selection but emerged spontaneously from the way that the physical world works. Defining a field as "the behaviour of a dynamic system that is extended in space," Goodwin (2001, 51–52) writes:

> A new dimension to fields is emerging from the study of chemical systems such as the Beloussov-Zhabotinsky reaction and the similarity of its spatial patterns to those of living systems. This is the emphasis on self-organization, the capacity of these fields to generate patterns spontaneously without any specific instructions telling them what to do, as in a genetic program. These systems produce something out of nothing . . . There is no plan, no blueprint, no instructions about the pattern that emerges. What exists in the field is a set of relationships among the components of the system such that the dynamically stable state into which it goes naturally—what mathematicians call the generic (typical) state of the field—has spatial and temporal pattern.

We need to explore the relationship between traditional Darwinism, with its emphasis on selection, and this new concept of self-organization. Let us do this through an example noted and cherished by an early Darwinian, Chauncey Wright, and also used by D'Arcy Thompson to illustrate the supposed priority of form over function. I refer to phyllotaxis, mentioned briefly in Chapter 7, the pattern of clockwise and anticlockwise spirals shown by many plants when identical elements are packed together. The sunflower is the classic case, but the phenomenon is widespread—look at the back of any pinecone or see the pattern as you tear apart a cauliflower. In fact, over 80 percent of the quarter-million higher plants show it in one form or another. So it is a major aspect of the botanical world, and its explanation is hardly of borderline significance.

Phyllotaxis occurs because leaves or petals are produced at the center (the "growing apex") of the flower and then are pushed outward. The leaves follow a spiral as they appear at the center (known as the "genetic spiral") and, if growth is constant, then the angle between successive leaves is going to be constant. But what about the spirals (known technically as "parastichies") that catch one's eye? It has long been realized that one can express phyllotaxis in mathematical form by means of a formula discovered by the thirteenth-century Italian mathematician Leonardo Fibonacci. He searched for a means of calculating the growth of the offspring of a pair of rabbits and came up with the series formed by adding together the previous two members of the series, starting with zero and one. The series thus being 0, 1, 1, 2, 3, 5, 8, 13, . . . or, more generally, $n_j = n_{j-1} + n_{j-2}$. Botanists have found that the order of generation numbers of leaves on parastichies, one set clockwise and one set counterclockwise, on a particular species of plant are always related by being consecutive numbers of the Fibonacci series.

Why this pattern? To the pragmatist and selection-enthusiast Chauncey Wright the answer was obvious. This kind of arrangement gives the best way of exposing each leaf to the light, without undue overlap from its fellows. With this end in view, the differences between the various phyllotactic arrangements are so minute as not really to matter that much. "To realize simply and purely the property of the most thorough distribution, the most complete exposure to flight and air around the stem, and the most ample elbow-room, or space for expansion in the bud, is to realize a property that exists separately only in abstraction, like a line without breadth" (Wright 1871, quoted in Gray 1881, 125). The formalists will have none of this. D'Arcy Thompson listed one objection after another. The differences between the arrangements are indeed significant, the teleological intent is something which "cannot commend itself to a plain student of physical science," there are all sorts of other ratios that would do the job as well, the plant could have taken other and better paths to exposing the leaves to sunlight, and much more. "We come then without more ado to the conclusion that while the Fibonacci series stares us in the face in the fir-cone, it does so for mathematical reasons; and its supposed usefulness, and the hypothesis of its introduction into plant-structure through natural selection, are matters which deserve no place in the plain study of botanical phenomena" (Thompson 1948, 953).

As a parting shot, Thompson accused the Darwinian of "harking back to a school of mystical idealism," which is somewhat ironic in light of what his successors now argue. When faced with phyllotaxy, Brian Goodwin in particular becomes practically Pythagorean in his numerology. He begins with the happy observation that the vulgar-fraction series formed from dividing consecutive members of the Fibonacci series homes in on the irrational number 0.618, which is in turn what the ancient Greeks called the Golden Mean or Golden Section—the ratio of the sides of a rectangle such that the rectangle left after removing the biggest possible square is of the same proportions as the original rectangle. This is just the beginning, for you can get the Golden Mean out of circles also, if you divide up the perimeter in an appropriate way. This yields the major angle of 137.5 degrees, and—wonder of mathematical wonders—this is just about the angle you tend to get with successive leaves on the genetic spiral. "So plants with spiral phyllotaxis tend to locate successive leaves at an angle that divides the circle of the meristem in the proportions of the Golden Section. Plants seem to know a lot about harmonious properties and architectural principles" (Goodwin 2001, 127). (The meristem is the growing tip of a plant. The connections, of course, are not arbitrary but follow mathematically from the properties of lattices, which is what we have here.)

Now we set ourselves up a little experiment, where a ferrofluid (a fluid with magnetic properties) is dropped slowly into the center of a polarized film of oil. The drops repel each other and move away from the center. If you do this sufficiently slowly, each drop is affected only by the previous drop, and you simply get a pattern of alternation. But when it is done more rapidly, wonderful things start to happen (Goodwin 2001, 127–128):

> As the rate of adding drops (equivalent to the rate of initiation of leaves in a meristem) is increased, a new drop experiences repulsive forces from more than one previous drop, and the pattern changes: the initial simple symmetry of the alternate mode gets broken, and a spiral pattern begins to appear. It takes a while for the system to settle on a steady pattern, the duration of this transient depending on the rate of adding drops. If this is rapid, so that there is strong interaction between drops, then a stable pattern emerges rapidly and successive drops quickly settle into a divergence angle of 137.5°, the spirals obeying the normal Fibonacci series.

And so once again we get self-organization, and just as one can get different patterns by altering the rates at which the oil drops, so also in plants the different patterns reflect simply the rates at which the plants grow and generate leaves. In short, "the frequency of the different phyllotactic patterns in nature may simply reflect the relative probabilities of the morphogenetic trajectories of the various forms and have little to do with natural selection" (p. 132). Or as Kauffman (1995, 151) puts it: "Like the snowflake and its sixfold symmetry, the pinecone and its phyllotaxis may be part of order for free."

Darwinians are not convinced. The formalists overlook the "obvious possibility" that "natural selection may universally favor close packing by phyllotaxis over alternative arrangements" (Reeve and Sherman 1993, 21). In a way, though, one suspects a certain amount of talking at cross-purposes here. Even the most ardent adaptationist must agree that phyllotaxis is governed by the kinds of mathematical formulas discussed above. There has to be some physical way for the plant to produce its spiral patterns. It would be very odd (if not impossible) were no mathematical formulas followed at all in development. And in fact, some evidence strongly suggests that they are indeed followed. G. J. Mitchison has set up models of development on the supposition that, in order to prevent crowding, a plant has an inhibitor at work when a leaf is formed. Only gradually does the effect of this inhibitor fall away, allowing a new leaf to form. Formally, the situation is not that different from the oil drop model and, as expected, one gets Fibonacci patterns. Indeed, by varying rates, one can show how Fibonacci ratios get changed and why. When the need for more parastichies arises, some bigger plants—sunflowers, for instance—switch ratios in the course of growth.

So, to go back to the language of constraints, phyllotaxis may well be produced because development goes down certain fixed physical channels. Hardly a great surprise, although working out the details is important and sometimes difficult. But from the Darwinian perspective, none of this matters one whit so long as the patterns produced are adaptive and therefore preserved through natural selection. Form does not preclude function. Or if the patterns as such are fixed, nothing too much matters from an adaptive viewpoint, so long as other plant features can be varied in order to maximize the efficiency of leaf exposure. And, in fact, the evidence suggests that this is so. Through computer simulations, Karl Niklas has shown that different phyllotactic patterns can influence the amount of light that plants intercept, but this can be compensated for by many other morphological features, for instance by varying the deflection ("tilt") angle of leaves. In other words, there is lots of room for adaptive excellence, whatever the constraints imposed by the mathematics of phyllotaxy. Indeed, Niklas prefers to speak of phyllotaxy as a limiting factor rather than something involving constraint—as the background in which adaptation is embedded, as that which makes adaptation possible. "The distinction between a 'constraint' and a limiting factor is important, because it reflects a measure of plasticity within the developmental repertoire." So to think that we have something here that threatens or denies the Darwinian approach is to misunderstand the nature and causes of the evolutionary process. Or, to put matters a little more moderately, discovering the details of phyllotaxis and its production shows that rules really exist in the nonliving world that must be followed by living organisms. But even if organisms cannot break the rules of chemistry or physics, they usually find ways to work within them to their own ends. It is all a bit like paying taxes.

Form or Function?

Supporters of form have come back in force. Generally, they have no desire to overthrow Darwinism as such, and just about all of them give natural selection some role in evolution. But many think that selection as a mechanism was much overrated and want to recognize other forces for the part they play. "Constraint" is the mantra of these new formalists. No formalist denies organized complexity in the organic world. Yet all argue that selection is not all-powerful and that such complexity as there is cannot be due entirely to selection. Responding to this challenge, Darwinians agree that much of molecular biology centering on development is incredibly exciting and at the front of the very best work being done on evolution in the past two decades. No one could have dreamed of the genetic homologies that seem now to be commonplace and at the very heart of the evolutionary process. All of this is surely transforming our thinking about descent with modification.

But—moving now from exposition to judgment—the extent to which this is a challenge to conventional Darwinism is entirely questionable. Selection obviously does have to work within physical and chemical limits and comes up against barriers—and all of this affects the end products. But the position of natural selection as the mechanism that winnows out the various solutions that (constrained) organisms offer up is still secure. Indeed, some would say that natural selection has been invigorated and made even healthier precisely because of the new emphasis on form, rather than despite it.

CHAPTER 12

FROM FUNCTION TO DESIGN

EVEN AS THE SCIENCE of evolution surged ahead, it had to fight for a place in the sun. By mid-century, molecular biology was beginning to soak up all of the funds and prestige, as well as attracting many of the brightest students away from organismic and evolutionary biology. In addition to doing their own work, evolutionists faced the task of persuading themselves and others that their field, though not physics or chemistry, was a real scientific program in its own right. As part of this persuasion, evolutionists had to demonstrate that their subject had moved beyond the sentiments expressed in the decades after the *Origin,* when evolution was in many respects no more than a secular religion, a vehicle for all sorts of moral and social claims.

In the terms of our division of the argument from design into the argument to complexity and the argument to design, the complaints of the formalists in the last chapter were of two kinds. Some, like Gould, argued that Darwinians are overly obsessed with the argument to complexity; in a more balanced approach there is no need to overuse natural selection as an alternative to the theological argument to design. Some, like D'Arcy Thompson, accepted the argument to complexity but argued that mathematics and physics can do as good a job (or better) than can natural selection in providing a scientific alternative to the argument to design. Now, in this chapter, we focus on the worry of Darwinians that by accepting a full-blooded scientific argument to complexity based on natural selection, they have somehow nevertheless failed to dispel the theological odor of this argument. We will look at those who have felt a need to purify the argument to complexity, so that no religious stench hangs over evolutionary theory.

A major part of this cleanup campaign has focused on the problem of function and purpose—a problem that remains with us to this day. Ernst Mayr, one of the leaders of the new evolutionism, was open about his concerns. First, with a philosophy of end-directness come all sorts of metaphysical, mystical, or supernatural notions that good scientists avoid in their work. Second, teleological thinking denies the applicability of the laws of the physical sciences throughout the natural world. Third, it seems to entail paradoxical consequences about causation, including such counter-intuitive notions as missing goals that could nevertheless influence events. Fourth, this kind of function-talk introduces an unacceptable anthropomorphism—the same old worry about words like "purpose," "intent," "design," or "contrivance" that Wallace had expressed to Darwin.

Let us begin with the first worry, using it as our entry into the debate.

The Vitalists

At the beginning of the twentieth century, a number of thinkers, dissatisfied with Darwinism—dissatisfied, in fact, with all current theories of evolution—suggested that life itself demands more than a purely naturalistic approach. They argued that one must transcend materialism and mechanism and appeal to "life forces" in some way. The German embryologist Hans Driesch was one such vitalist; he argued that we need to invoke something that he termed "entelechie." Another was the French philosopher Henri Bergson, who spoke to a need obviously felt by many. Bergson was enthusiastic about evolution and committed to its program, but he nevertheless concluded that no then-existing mechanistic theory of evolution had succeeded in explaining all of the pertinent facts.

In particular, following earlier critics of Darwin, Bergson made much of the problem that was later to worry Sewall Wright, the problem of complexity, and especially the difficulty of producing, through purely naturalistic means, something as complicated as the eye, which seems to have evolved independently in many separate lines. "An accidental variation, however minute, implies the working of a great number of small physical and chemical causes. An accumulation of accidental variations, such as would be necessary to produce a complex structure, requires therefore the concurrence of an almost infinite number of infinitesimal causes. Why should these causes, entirely accidental, recur the same, and in the same order, at different points of space and time?" (Bergson 1911, 59–60).

According to Bergson, we need something more than a theory based purely on a mechanism like natural selection. But we need more than just something more: a mechanistic view of any kind will not do. It destroys the holistic view that organisms require. "The real whole might well be, we conceive, an indivisible continuity. The systems we cut out within it would, properly speaking, not then be *parts* at all; they would [be] *partial views* of the whole. And, with these partial views put end to end, you will not make even a beginning of the reconstruction of the whole." This decompositional approach spells the end of trying to understand the essence of life. "Analysis will undoubtedly resolve the process of organic creation into an ever-growing number of physicochemical phenomena, and chemists and physicists will have to do, of course, with nothing but these. But it does not follow that chemistry and physics will ever give us the key to life" (Bergson 1911, 32–33).

What Bergson sought as an alternative was something that would give direction to evolution. He was explicit in his end-directed vision. At the beginning of *Creative Evolution*, his major work on biology, he wrote: "The history of the evolution of life, incomplete as it yet is, already reveals to us how the intellect has been formed, by an uninterrupted progress, along a line which ascends through the vertebrate series up to man" (p. ix). However, Bergson pulled back from what he called a fully finalistic view, where everything is simply predetermined before one begins. The history of life shows far too much variation and randomness for that. One needs a kind of creative force, end-directed, that keeps pushing upward toward consciousness and intelligence, and ultimately humankind.

Bergson found this creative power in the *élan vital*, or life force, possessed by all living things. "This impetus, sustained right along the lines of evolution among which it gets divided, is the fundamental cause of variations, at least of those that are regularly passed on, that accumulate and create new species" (pp. 92–93). Normally, variations take organic groups apart and away from one another. But, Bergson explained, at times the impetus takes the variations along similar lines, creating a kind of parallel or identical evolution, as in the case of eyes. Not that one should think that the *élan vital* predetermines everything. It works like consciousness, as organisms decide on the best path to be taken and then try to achieve it. There is direction, and the end influences the course of events, but it does so through something like consciousness. Defining life as "a tendency to act on inert matter," Bergson wrote that "the direction of this action is not predetermined; hence the unforeseeable variety of forms which life, in evolving, sows along its path. But this action always presents, to some extent, the character of contingency; it implies at least a rudiment of choice. Now a choice involves the anticipatory idea of several possible actions. Possibilities of action must therefore be marked out for the living being before the action itself" (pp. 101–102). Apparently, sight is just such a possibility of action, and that is why complex eyes have evolved several different times.

Bergson's ideas were highly influential among those who would later professionalize evolutionary studies. Sewall Wright's worries about the ability of evolution to explain complexity stem straight from his youthful enthusiasm for Bergson. Julian Huxley's first book, *The Individual in the Animal Kingdom,* was explicitly Bergsonian. Theodosius Dobzhansky was a third major evolutionist who responded sympathetically to Bergson's ideas. But for all this, such ideas were greatly out of tune with modern science, and this was recognized even by those who felt his influence most strongly. In *Evolution: The Modern Synthesis,* Julian Huxley wrote: "Bergson's *élan vital* can serve as a symbolic description of the thrust of life during its evolution, but not as a scientific explanation. To read *L'Evolution Créatice* is to realize that Bergson was a writer of great vision but with little biological understanding, a good poet but a bad scientist" (Huxley 1942, 457–458).

The problem with the *élan vital* was not so much that it was unseen or directly unknowable. Science is full of unseen and directly unknowable entities—electrons, for a start. The problem was that the *élan* (unlike the electron) was not embedded in any laws and was ultimately useless for prediction or unification or any of the other epistemic demands that one makes of the unseen entities of science. Electrons cannot be seen, but they perform in predictable ways that can be studied in the laboratory. With the *élan,* on the other hand, one could do just as much evolutionary biology without it as one could do with it. It gave the impression of explanatory power, but it had no substantive payoff. As the paleontogist George Gaylord Simpson (1949, 125) wrote: "Granting, as any reasonable person must, that there is an important difference between life and non-life, you may, if you wish, call the different behavior of matter in life 'vitalistic,' but this accomplishes nothing and means nothing that was not already obvious."

This was true in the days before the DNA molecule and the double helix was discovered, and it was even more the case afterward. In what sense did the *élan vital* help biologists crack the genetic code or work out the development of the organism from the information of the macromolecules of heredity to the finished adult? More than this, the *élan vital* seemed to entail some highly counterintuitive ideas. For all that Bergson claimed that he did not want to ascribe real consciousness to all living matter, that was precisely the direction in which he was pointing. If the trilobite, say, or the plant did not have consciousness of some kind, then how could it choose to go in one direction rather than another?

The *élan vital* and other vitalistic notions would not do. The turn-of-the-century English Darwinian Edward B. Poulton, to his credit, was scathing on the subject. And with the coming of neo-Darwinism in the 1930s and 1940s, a bright light—pertinent, explanatory light—was thrown on the kinds of problems that worried people like Bergson. Genetics put selection-based evolutionary theory on a firm foundation that it did not enjoy when Bergson was writing. Much had been learned about the nature of variation, and—thanks to theorists like Fisher and practitioners like Dobzhansky—the neo-Darwinians had a better grasp of how purely random mutations might nevertheless become the basis of highly complex characteristics. Work on adaptation during the last fifty years and more shows that many of Bergson's fears were perhaps understandable but nevertheless ungrounded. Under the forces of natural selection working on nondirected variations, complex structures such as the eye can indeed evolve and have in fact evolved many times.

Goal-Directed Systems

During World War II engineers succeeded in building machines that could respond flexibly to moving targets—torpedoes that did not simply go in straight lines but tracked the ships at which they were directed. This led to speculations about end-directed behavior by a number of theoretical biologists and like-minded philosophers, especially those from the logical empiricist school. These people—much influenced by successes in the physical sciences—hoped to articulate the nature of science from an empiricist perspective using the formal insights of modern logic and mathematics. Naturally enough, they were entirely unsympathetic to vitalism or finalism of any kind and yearned in theory if not in fact for the "reduction" of the biological sciences (including evolutionary theory) to the physical sciences.

At the ontological level, this meant an ultimate understanding of living things only in terms of the material entities like molecules of which they were made. At the theoretical level, this meant deduction of the ideas and principles of biology from those of the physical sciences. Clearly, if this program (theoretical reduction in particular) was to be effected, then in some way the seeming purposefulness of evolution must be explained away or explained in terms of the non-end-directed notions of physics and chemistry.

The key idea behind the logical empiricists' attack was feedback, meaning a system where, as the target moves or changes, the homing device—a torpedo headed toward a ship, for instance—gets back fresh information from the target and adjusts itself accordingly to achieve the desired goal. In other words, as the American philosopher Ernest Nagel stressed, when something goes wrong in the system, something in the system fixes it. In such a goal-directed or directively organized system, it is not enough that the goal is attained or attainable but that the system have some flexibility to respond to disruptive variations and to bring itself back on track (toward the goal) again.

Let us grant what is surely true, that there does seem to be something purposeful about a goal-directed system. A more-than-chance connection exists between torpedoes and the sides of ships, even though the one does not always hit against the other. Let us also grant that goal-directed systems occur in the living world. The human body's capability to maintain a constant temperature, thanks to sweating and shivering, is a paradigm example of feedback at work in a goal-directed system. And let us finally grant (what again is surely true) that nothing inherently supernatural or nonphysical is required by such systems, and nothing about them is illogical. No consciousness is involved, and no goal objects are missing. Sometimes even the best missiles fail to hit their targets; sometimes people freeze to death or get fatally over-heated. But normally in the human body, as in naval warfare, feedback and what Nagel would call responsive adaptive variation allow the system to adjust to new circumstances and new information. Granted all that, what does it imply for with the kinds of seeming goal-directedness we encounter in evolutionary biology—the behavior of those dunnocks and cuckoos, for example?

Nagel (1961, 403) argued that a very close and direct relationship exists between evolution and goal-directed systems:

> Consider a typical teleological statement in biology, for example, "The function of chlorophyll in plants is to enable plants to perform photosynthesis (i.e., to form starch from carbon dioxide and water in the presence of sunlight)." This statement accounts for the presence of chlorophyll (a certain substance A) in plants (in every member S of a class of systems, each of which has a certain organization C of component parts and processes). It does so by declaring that, when a plant is provided with water, carbon dioxide, and sunlight (when S is placed in a certain "internal" and "external" environment E), it manufactures starch (a certain process P takes place yielding a definite product or outcome) only if the plant contains chlorophyll.

In this passage Nagel suggested that a teleological statement is an argument—a condensed argument but an argument nevertheless. "When supplied with water, carbon dioxide, and sunlight, plants produce starch; if plants have no chlorophyll, even though they have water, carbon dioxide, and sunlight, they do not manufacture starch; hence, plants contain chlorophyll" (p. 403). But, as Nagel recognized, this is not enough to explain what happens in plants. Consider Boyle's law, that a volume of a gas at constant temperature varies inversely with respect to pressure. This law makes no claims about purpose, and if we tried to assign a purpose to it we would be laughed out of court. "The function of varying pressure in a gas at constant temperature is to produce an inversely varying volume of the gas." Or how about: "Every gas at constant temperature under a variable pressure alters its volume in order to keep the product of the pressure and the volume constant"? If we tried this, we would be considered slightly crazy. As Nagel puts it, "Most physicists would undoubtedly regard these formulations as preposterous, and at best as misleading.

Accordingly, if no teleological statement can correctly translate a law of physics, the contention that for every teleological statement a logically equivalent nonteleological one can be constructed seems hardly tenable."

Here is where the notion of goal-directedness comes in. For Nagel, the difference between a statement that can bear a functional reading and a statement that cannot bear such a reading is that the former (and the former alone) refers to a system that is directively organized. In other words, a functional statement implicitly acknowledges that biological systems have a goal-directed nature. To speak of the function of chlorophyll as performing photosynthesis is to acknowledge the fact that plants are goal-directed systems. And this explains why teleological language seems out of place in the physical sciences (p. 421):

> A teleological version of Boyle's law appears strange and unacceptable because such a formulation would usually be construed as resting on the assumption that a gas enclosed in a volume is a directively organized system, in contradiction to the normally accepted assumption that a volume of gas is not such a system . . . In a sense, therefore, a teleological explanation does connote more than does its prima facie equivalent nonteleological translation. For the former presupposes, while the latter normally does not, that the system under consideration in the explanation is directively organized.

Nagel drew his conclusion. There are no special forces or entities powering living things. Organisms are as much material objects as are inert chemicals. Ontologically, therefore, organisms can be thought of as at one with, or reduced to, the entities of physics and chemistry. And more than this. What drives or governs organisms are processes obeying normal—that is, physical and chemical—laws. Theoretically, therefore, the reduction of biology to physics and chemistry is a viable project. The difference between evolutionary biology and physics and chemistry is one of emphasis and reference, nothing more.

Causes and Capacities

One thing that never troubled logical empiricists like Nagel was an unduly thorough acquaintance with biology, with evolutionary theory and practice in particular. Protestations notwithstanding—and there were not too many of these at mid-century—if it was not physics and chemistry, then it was not real science, according to this school.

But already, approaches like Nagel's were under attack. The geneticist C. H. Waddington (1957) put his finger on the main problem, when he complained that the situation being characterized as a goal-directed system is one that biologists would call a case of adaptability. Humans are adaptable, especially in their ability to maintain a constant internal body temperature despite a range of external environmental temperatures. However, the kind of case where we would normally use functional language focuses not on adaptability but on adaptation. Instances of adaptability are usually (probably always) going to involve adaptations. Keeping the blood temperature constant is undoubtedly of adaptive significance. And organisms are usually (probably always) in some way adaptable. But adaptation does not always require directive organization. An adaptation may be very rigid in its behavior or function. The teeth of ungulates, for instance, are well adapted for the diet consumed by their possessors—coarse vegetable matter with little nutritional content, like grass. But if the climate changed or if the availability of the usual feed material vanished, ungulates have no guarantee that their teeth could be used for other foodstuffs or that their teeth could change in a desirable direction.

Waddington's general conclusion therefore was that goal-directedness, however attractive on the surface, is essentially irrelevant to the question of whether adaptations have a purpose. As it happens I shall suggest that this conclusion (drawn by me as much as by anyone) is perhaps a little hasty, and that goal-directedness does in fact have some place in our analyses. But it is clear that the main focus should have been (as George Williams was telling all who would hear) on adaptation. "When a biologist asks a what-for-question of some organic feature he postulates that it is an adaptation. (In so doing he may, of course, be wrong.) He then tries to discover the trait's 'adaptive significance,' that is, why it was selected for" (Williams 1966, 91). But while this may all be true (and I think it is), in a way this only takes us part of the route.

In the 1970s, two much-discussed alternative analyses were proposed that purportedly covered both biology and the world of human activity, the two domains where we find end-directed thought. The first proposal, credited to the philosopher Larry Wright, was known as the causal (or etiological) analysis. The proposal was very Kantian in spirit, for it involved a two-way causal connection. "A does B. A exists because it does B." In Kant's language: "I would say that a thing exists as physical end *if it is* (though in a double sense) *both cause and effect of itself*" (Kant 1790, 18). In Wright's language: "When we say that Z is the function of X, we are not only saying that X is there because it does Z, we are also saying that Z is (or happens as) a result or consequence of X's being there." So, in terms of familiar examples: "Not only is chlorophyll in plants *because* it allows them to perform photosynthesis, photosynthesis is a *consequence* of the chlorophyll's being there. Not only is the valve-adjusting screw there *because* it allows the clearance to be easily adjusted, the possibility of easy adjustment is a *consequence* of the screw's being there" (Wright 1973, 160). Notice that all of this has to be understood in a generic sense, otherwise we run into missing-goal-object problems and the like. X does not always have to do Z in every case—it may be that X does Z on only a few occasions—but it must do it sometimes, and these must matter.

The second proposal, credited to the philosopher Robert Cummins, was known as the capacity analysis. Here the emphasis was on something's doing something or having the ability or capacity to do something within an overall system. Let us suppose we have a system—an organism or a complex artifact like an electrical system—and let us suppose that it does something, like survive and reproduce or produce a kind of electrical current. Now let us suppose that we have a component within that system, like a heart or a particular sort of battery, and that that component does something, like pump the blood or produce power, which is needed for the overall working of the system. Cummins suggested that functional language is appropriate under these circumstances, because the component does something within the system that helps the system to do its overall task. More particularly, the component has the function of doing something within the system which enables the system to fulfill its purpose. So we would say that the heart functions as a pump in humans because and precisely because, against a background ability to survive and reproduce (that is, against the overall working of the body), the heart is needed for and contributes to that ability. If the heart as pump did not contribute to survival and reproduction, we would not use functional language. Insofar as the sounds a heart makes while pumping make no difference to the success of the body, the function of the heart is not to produce such sounds.

Showing how his analysis applied indifferently to both human-made situations and to biology, Cummins wrote first about schematic diagrams in electronics. "Since each symbol represents any physical object whatever having a certain capacity, a schematic diagram of a complex device constitutes an analysis of the electronic capacities of the device as a whole into the capacities of its components. Such an analysis allows us to explain how the device as a whole exercises the analyzed capacity, for it allows us to see exercises of the analyzed capacity as programmed exercise of the analyzing capacities." Then, turning to biology, he pointed to an exactly similar situation. "The biologically significant capacities of an entire organism are explained by analyzing the organism into a number of 'systems'—the circulatory system, the digestive system, the nervous system, etc. each of which has its characteristic capacities. These capacities are in turn analyzed into capacities of component organs and structures. Ideally, this strategy is pressed until physiology takes over—i.e., until the analyzing capacities are amenable to the instantiation strategy."

Cummins went on to draw the analogical strings even more tightly. "We can easily imagine biologists expressing their analyses in a form analogous to the schematic diagrams of electrical engineering, with special symbols for pumps, pipes, filters, and so on. Indeed, analyses of even simple cognitive capacities are typically expressed in flow-charts or programs, forms designed specifically to represent analyses of information-processing capabilities generally" (Cummins 1975, 188–189).

Analysis

Now, to what extent do these two proposals carry us forward in our quest to understand goal-directedness in the living world? I am going to say little about Cummins's suggestion, which is not so much wrong as it is not terribly helpful to us in our inquiry here. Cummins himself has taken almost perverse pride in the fact that his thinking about function has been divorced from evolutionary questions. Cummins was aiming at such a level of generality that he really was not trying to analyze function specifically as it occurs in evolutionary biology. He was using "function" simply in the sense of "How does it work?" and had no interest in overall biological benefits or ends. The point for us, however, is that "function" comes in the context of "adaptation" and that is precisely what we want to unpack, not ignore or downgrade. In fact, I believe that Cummins's analysis does highlight something significant in the biological understanding of function, namely, the parallel between the use of function-language in both biological and artifactual systems. Although this perhaps accounts for the continued popularity of his account, this for him was not the essence of the situation.

I must admit that some morphologists and their philosophical sympathizers have suggested that a capacity analysis is indeed precisely what happens in that branch of science. The morphologist is concerned with working out how something "functions" and as such has little interest in overall evolutionary considerations. The philosopher Ron Amundson and biologist George Lauder have written that "Cummins's account is of special interest because of its close match to the concepts of function used within functional anatomy. His emphasis on causal capacities of components and the absence of essential reference to overall systemic goals is shared by the anatomists" (Amundson and Lauder 1994, 448). And they continue: "Concepts involving biological importance, selective value, and (especially) selective history (and therefore Darwinian adaptation), are all at higher and more inferential levels of analysis than that of anatomical function" (p. 449). Which is all true. In fact, what they claim is borne out by history. The professional, evolutionary morphology of the post-*Origin* decades owed far more to the nonevolutionary *Naturphilosophen* (and their students, like Louis Agassiz) than it did to Darwin. We would expect that morphology, even today, could be carried on in a fashion where evolution—Darwinism especially—is essentially irrelevant, where one is finding out how things work and relate, how things function within a system. But today we live in a Darwinian world, a world of natural selection which insists on a higher level of analysis, a Darwinian level of analysis beyond pure morphology.

Turning now to Wright's analysis—or, as I would prefer to say, the Kant/Wright analysis—it does surely carry us some way forward in our investigation. It does indeed tie together the human and the biological realms, showing that distinctive in both cases is the rather curious temporal reversal that is characteristic of teleological explanations or accounts. The function of the bell is to tell you that the meat is cooked. The bell is there because it tells you that the meat is cooked. You are told that the meat is cooked because the bell is there. The function of the eye is to focus light. The lens is there because it focuses the light. The light is focused because the lens is there. The analysis as such does not tell you how to wriggle out of the temporal reversal without invoking unwanted entities like causes working backward, but it is not too difficult to see how the different situations lead to (demand) the same kind of analysis, without any need of peculiar causes working out of the future or some such thing. In the human case, the thought about the meat's ready state leads to the insertion of the timer with a bell. In the biological case, we are pointed to—what is very much the case with natural selection—a kind of cyclical cause and effect situation. The lens leads to successful focusing, which, because it aids its possessors, leads to the production of more lenses of a similar kind. X leads to Z, which in turn leads to the production of more Xs, and so the process keeps going. Kant again: "A tree produces . . . another tree according to a familiar law of nature. But the tree which it produces is of the same genus. Hence, in its *genus*, it produces itself. In the genus, now as effect, now as cause, continually generated from itself and likewise generating itself, it preserves itself generically" (Kant 1790, 18).

However, something is still not quite right here, or at the very least something is missing. Think of a natural, nonorganic, cyclical situation: let us say the cycle that leads to rainfall. The rain falls on the mountain, it is carried by rivers down to the sea, the water is evaporated by the sun, forming new rain clouds, which discharge their content as they float over land and up to the mountains. The river is there because it produces or conveys water to form new rain clouds. If there were no water going into the sea, eventually the river would dry up. The rain clouds are a result of the river's being there. But we would hardly want to say that the function of the river is to produce rain clouds. Or if we did say that, it would only be in Cummins's nonteleological sense.

The problem is that the river does not exist in order to produce rain clouds. While the river may be a required condition of the production of rain clouds, it is hardly needed in the sense of desired or wanted. But the timer really does exist in order to produce perfectly cooked meat. It really is needed in the sense of desired or wanted. Likewise, the lens does exist in order to produce good vision—it really is needed in the sense of desired or wanted. And this surely points to what is absent. We have forgotten our Plato: purpose occurs when *values* are at stake: "If anyone wanted to discover the cause of anything, how it came into being or perished or existed, he simply needed to discover what kind of existence was *best* for it, or what it was best that it should do or have done to it" (*Phaedo*, 97 b–c). Perfectly cooked meat is considered a good thing—it is pleasant to the taste and good for the digestion. Well-focused lenses are considered a good thing—they help with vision and hence with survival and reproduction. The end is not necessarily a human end—we use functional language when speaking of parasites—but it does exist for someone or something. Rivers are not a good thing or a bad thing, at least not outside the world of humans and other organisms. They just run down to the sea.

A system where functional language is appropriate is a value-impregnated system. In both the human case and the biological case we are concerned with ends that are valued or desirable. Which tells us at once why this point has been so difficult for philosophers to grasp, for a major legacy of logical empiricism (and its predecessor, logical positivism) is that science is taken to be value-free. In some sense, science supposedly tells us like it is, unlike philosophy or religion or other subjects that tell us like it should be or like we would want it to be or some such thing. Science is objective, values are subjective, hence science (good science, that is) has no values, so said the logical empiricists. Hence any analysis of teleological thought which is going to preserve the integrity and worth of evolutionary biology must be one that eliminates or avoids all value-talk. What we get is the kind of emaciated analysis offered by Wright.

Design

Values come in through ends. Things are judged useful, and so their parts are said to serve various functions as they contribute to these ends. The bell functions to tell us that the meat is perfectly cooked. What are these things that are judged useful? They are artifacts—Paley's watch, the telescope, the meat timer. Functional talk makes sense in a situation where humans are making things with certain ends and the functions contribute to those ends. This is why a temporal reversal in explanation is possible, as one tries to see or understand why things exist—they exist (as Wright correctly stresses) for certain ends, and the point is (as Wright fails to see or mention) these ends are valued or desirable.

As Darwinians have already flagged for us, something exactly analogous—or metaphorical—occurs in the biological case. We are faced with what we have been calling, in as neutral a way as possible, "organized complexity"—although from the beginning it was admitted that this is not very neutral, and at a minimum one should rather speak of "seemingly organized complexity." Whatever you call it, this complexity allows for and indeed calls for understanding in human terms of intentionality, of purpose, of *design*. Organized complexity is artifactual. That was the whole point of natural theology—the argument to design—and the moment and place where Darwin took over. Whether or not organisms really are designed, thanks to natural selection they (or rather, they inasmuch as they are adaptive or adapted) seem as if designed (for the ends of survival and reproduction). We may no longer be thinking of a literal designer up there in the sky, but the mode of understanding persists. For the natural theologian, the heart is literally designed by God—metaphorically, we compare it to a pump made by humans. For the Darwinian, the heart is made through natural selection, but we continue, metaphorically, to understand it as a pump made by humans.

When you use a metaphor, you are taking an idea from one field and applying it to another field—so the second field is being seen through the lens of the metaphor. Strictly speaking the metaphorical term is illicit, false, when applied to the second field, but the similarities and juxtaposition—and the shock of the move itself—stimulate you to see the similarities, and perhaps even more than there were at first. So a new truth emerges, one that is metaphorical at first but clearly understandable, and moves toward the literal. This is a major difference between a metaphor and a mere analogy—similarity is involved in both, but analogies stay with the original literal truth, whereas metaphors make you rethink what you are now prepared to consider true. And the shock of the metaphor pushes you to further thought.

As we shall see, a metaphor has a heuristic function. That is why

Thomas Kuhn, in his *Structure of Scientific Revolutions,* when he wanted to stress that scientific change is abrupt and shocking and moves to new and fertile ways of thinking (new "paradigms," in his language), insisted that his theory was one intimately connected with metaphorical change and innovation. To regard the heart as the pump is to think about it in a whole new way, leading you to ask all sorts of questions about values and rates of flow that you would not otherwise ask. Without the heart's being a pump, these questions make no sense.

The metaphor of design, with the organism as artifact, is at the heart of Darwinian evolutionary biology. But let us not just take someone's word for this. Let us have before us a paradigm case of adaptationism, an adaptive explanation that would be insisted upon by a Darwinian evolutionist. A case where functional language—final-cause talk—would not simply be allowed but insisted upon. Consider the trilobites, those well-known, long-extinct, marine invertebrates whose position today is occupied by crabs and like creatures. Their eyes were very complex, with multiple lenses like one of today's insects. If we cut through one such lens, we find that in fact it seems to have been made of two lenses, divided by a distinctively shaped barrier. Why should the lenses have had these particular shapes and not others? Could this be chance? Apparently not. With such complex lenses, it seems that the creatures avoided spherical aberration: light being focused differently according to the different wavelengths of the component light.

Is this really so? By way of proof, students of the trilobites offer the diagrams worked out by the seventeenth-century physicists Descartes and Huygens, exactly the forms discovered by the trilobites. Such a coincidence is no chance but evidence that something is afoot. More precisely, the coincidence points to a design-like feature of the organic world, which Darwinians would take as evidence of adaptation—features possessed by organisms because they help in the struggle to survive and reproduce—and, as Darwinians, this adaptation would be explained as the end result of a selective process.

Design. That is why teleological language is thought appropriate. Adaptations are artifact-like. We talk of ends when we come to human-made objects, so analogously (or metaphorically) we talk of ends when we come to adaptations. If the heart were not like a (human-made) pump, teleological language would never arise, but it is and so it does. If the trilobite eye's lens were not like a human-designed lens, made for a particular purpose, teleological language would never arise, but it is and so it does. Ricardo Levi-Setti (1993, 54) put it this way: "Of course, the laws of physics existed prior to their discovery by man. And we shouldn't perhaps be too surprised that the drive to optimize biological function—one of the fundamental evolutionary forces in all biological organisms—caused trilobites to follow physical laws to the fullest possible extent in their development of visual systems. The real surprise should not be that they did construct eyes that work according to the laws of physics, *but that they did it with such ingenuity.*"

The paleontologist-historian Martin Rudwick makes precisely this point. He asks why we might think the function of the wings of an extinct organism like a pterodactyl could be for flying, but only gliding. Why would we say that "their function is flight through gliding"? Rudwick agrees that, of course, we are going to work by analogy from flying organisms today. But, ultimately, we must go beyond this. "From our knowledge of natural and artificial aerofoils, and of the structural requirements of their successful operation, we conclude that the pterodactyl forelimb would have been physically capable of functioning as an aerofoil. From our knowledge of the energy requirements for powered flight and of the energy output of vertebrate muscle, we conclude that it would not have been capable of functioning as a flapping wing for powered flight." Even if there were no gliding animals today, we could still work all of this out. "All we need, ideally, is a knowledge of the operational principles involved in all actual or conceivable flight mechanisms possible in this universe. Consequently the range of our functional inferences about fossils is limited not by the range of adaptations that happen to be possessed by organisms at present alive, but by the range of our understanding of the problems of engineering." And obviously here we are into artifacts/adaptation/function. "An analysis of fossil structures in the light of operational principles is thus, I believe, fundamental and unavoidable in any inference about their possible functions or adaptive significance. This involves the limited 'teleology' that is inherent in any description of a machine as a machine" (Rudwick 1964).

This way of thinking and procedure is precisely what we have seen in our chapters on modern-day scientific thinking, most especially in the work on the stegosaurus back plates. To rephrase: "From our knowledge of natural and artificial cooling-systems fins, and of the structural requirements of their successful operation, we conclude that the Stegosaurus back plate would have been physically capable of functioning as a cooling fin. From our knowledge of the stress requirements for defence or fighting and of the structural nature of the back plate, we conclude that it would not have been capable of functioning as a tool of defence or attack."

Let us sum up. There is no reason to think that biology calls for special life forces over and above the usual processes of physics and chemistry. Nor is there reason to think that biology is little more than complicated physics and chemistry, distinguished only inasmuch as it makes implicit reference to purely mechanical, goal-directed systems. There does seem to be something distinctive about biological understanding—something having to do with purposes and ends in evolution. It seems, at one level, to lie in the fact that in evolutionary biology we are dealing with peculiarly connected, causal relationships. We have "final causes." One set of things causes another set of things, but the first set in turn seems to exist *because* it brings about the second set. But more: the second set of things is in some sense desirable, to be valued. Whether this appeal to values violates the norms of good science is a topic to which we must return. However, if we can agree now that such an appeal to values pertains, then we start to see that design-type thinking occurs precisely because it is in the manufacture of artifacts that valued ends become pertinent and important. We humans make objects for certain desired consequences, and (as in biology) we understand the parts of these artifacts in terms of their functions for the ends we desire. Both history and present Darwinian evolutionary practice have shown us that this kind of design-type thinking is involved in the adaptationist paradigm. We treat organisms—the parts at least—as if they were manufactured, as if they were designed, and then we try to work out their functions. End-directed thinking—teleological thinking—is appropriate in biology because, and only because, organisms seem as if they were manufactured, as if they had been created by an intelligence and put to work.

The argument to organized complexity, the argument to design-like complexity—truly, this is what is at the center of Darwinian evolutionary biology. If one is thinking just in terms of science, then it is virtually tautological that Darwinism holds the key to our solution. The design of organisms is to be understood in terms of their survival and reproduction, as Darwin insisted. And the strange causal connections come out because of Darwin. Something is of value because it leads to the end of survival and reproduction, but this survival and reproduction are in turn the reason why it exists. A leads to B, but then B in turn leads to A. There is no backward causation here, nor is there (as Bergson implied) some kind of intentionality. A does not have B in mind, nor does a creator have B in mind when A is made. It is rather a cyclical situation, where the first leads to the second and then back to the first. Darwinism does not have design built in as a premise, but the design emerges as Darwinism does its work and some organisms get naturally selected over others.

We have spoken to Mayr's second and third worries (about the applicability of the laws of the physical sciences and the avoidance of paradoxes like missing goals). What about our philosophers and earlier thinkers? Are we in their tradition? Plato put us on track from the beginning, by thinking in terms of design and by stressing values. He did not focus on the distinctiveness of biological thinking, but this is at least implicit in Aristotle, who saw fully that understanding biological entities requires reference to final causes. We may not accept Aristotle's answer (which many think of as being in some way vitalistic), but we applaud his insistence that any solution to the understanding of biological entities must be in their own terms rather than by appeal to extra-scientific entities. Aristotle was important for his internalist approach as opposed to the externalist approach of Plato.

The Kantian nature of the so-called etiological approach has been noted already—the connection between cause and effect and back to cause again—and Cuvier of course had a direct influence on Darwin that is reflected in our analysis. Our emphasis has been on the integrated working, the functioning, of the organism, as something that is distinctive of the living world. Finally, no argument is needed to show the importance of natural theology—looking back to Plato in the insistence that the world of organisms is as if designed, and looking forward to

Darwin and through him to us, in stressing adaptation for ends and the analogy with artifacts. The living world is seen as if it were the product of intelligence.

The pre-Darwinian philosophers saw the problems. They did not have the naturalistic answer that Darwin gave to us, and for this reason their solutions could be only partial. Kant saw the causal connections but did not realize that these lead to selection and through this to design-like effects. The natural theologians downplayed or had difficulties with the nonadaptive aspects of life (such as homology). But the positive overrides the negative. Where we are today fits very comfortably with the Western tradition of thinking about organisms and final causes. Ours is an evolution from the past rather than anything radically new. And that in itself is a comforting conclusion.

CHAPTER 13

DESIGN AS METAPHOR

NOW THAT THINGS have been spelled out, we see that there is nothing very mysterious about purpose in evolution. At the heart of modern evolutionary biology is the metaphor of design, and for this reason function-talk is appropriate. Organisms give the appearance of being designed, and thanks to Charles Darwin's discovery of natural selection we know why this is true. Natural selection produces artifact-like features, not by chance but because if they were not artifact-like they would not work and serve their possessors' needs.

Still, is it a concern that we have a metaphor here, a human-based metaphor? Are we not being unduly anthropomorphic? Remember Ernst Mayr's fourth worry: "The use of terms like purposive or goal-directed seemed to imply the transfer of human qualities, such as intent, purpose, planning, deliberation, or consciousness, to organic structures and to subhuman forms of life" (Mayr 1988, 40). Well, yes it does! But as Darwin pointed out, we use metaphors all the time in science, and many are based directly on human emotions and actions. Force, pressure, attraction, repulsion, work, charm, resistance are just a few of these many metaphors. Without metaphors of this kind, science would grind to a halt (to use a well-worn metaphor!). Darwin himself wrote: "The term 'natural selection' is in some respects a bad one, as it seems to imply conscious choice; but this will be disregarded after a little familiarity. No one objects to chemists speaking of 'elective affinity;' and certainly an acid has no more choice in combining with a base, than the conditions of life have in determining whether or not a new form be selected or preserved" (Darwin 1868, 1.6).

Frankly, I am not sure that Darwin ever entirely convinced himself on this issue. He kept scratching away at it. Certainly, he did not convince all of his contemporaries or subsequent evolutionists. The design metaphor—its putative existence and significance—is still a matter of much debate, so let us look at some of the issues.

The Metaphor of Design

Start with the most basic question. Does evolutionary biology really have at its heart the metaphor of design? Do evolutionists truly regard organisms as artifactual? Well, yes: we have the backing of history, we have the words of evolutionists themselves, and we have the examples of their work (remember the trilobite eye). Yet, some people, for instance the philosopher Colin Allen and the biologist Marc Bekoff, worry that there are dis-analogies between artifacts and organisms. Hence, they deny that we have a true metaphor. "Functional claims in biology are fully grounded in natural selection, and are not derivative of psychological notions such as design, intention, and purpose" (Allen and Bekoff 1995, 612). In the human case, they write: "Many people use natural objects (driftwood, seashells, heads of game animals, etc.) for decorating rooms and buildings. These objects are clearly not designed for that purpose (although they are presumably placed in strategic locations by design, in the sense of intent design). A rock on a desk may function as a paperweight, but unless the rock has had a flat base chiseled into it, or other similar modification, it is not appropriate to say that this object was designed for the purpose of holding down papers." Hence, "function does not entail design for that function" (p. 614).

This is not a strong point, although it does hint at important facts about the use of concepts and metaphors. With respect to the use of rocks and like objects for certain human functions, while the rocks themselves are not necessarily designed, they are put by design in certain places in particular ways, after making choices (we would not choose a ten-ton rock for a paperweight). At the very least we have borderline cases, which, outside mathematics, is what we encounter in real life all of the time anyway. Whether we think that a rock used as a paperweight is an instance of design is just such a borderline case. As are the kinds of borderline cases that we find in biology.

Suppose a characteristic has been produced by selection for one particular end—it is designed for this end and has such a function. Suppose now that the characteristic starts to get exploited for other ends. Bones may have started their evolutionary careers as calcium banks, and only later were used for the rigid supports of vertebrates. Does one want to say that the bones were designed as banks and have (and can only have) the function of banks? Or does one want to allow that the bones may have now a new function without necessarily having been designed for such an end? I have a geranium planter in my back garden which started life as a beer barrel. It changed functions, and, whether you tie design into its every moment or whether you say that it has a function for which it was not designed, the change does not negate a connection between function and design. By using the barrel as a planter, I am starting to change not just the function but the design—especially when I paint the barrel in bright colors. By using the bones as supports, nature changed not just the function but the design of bones—especially when selection started to work on the support virtues of bones.

The whole point about analogies and metaphors is they work only *because* of differences, as well as similarities, between the original object and that being compared metaphorically; what metaphors add is a certain looseness or open-endedness. "The moon is the night's answer to the sun." We know roughly what this means. The sun by day and the moon by night. But how are we to regard the moon? As warm and friendly like the sun, or as the exact opposite, as cold and hostile? The matter is open, and in this case depends on many factors, including the inflection of the person who makes the statement—is it being said supportively or in a rather cold and cynical fashion? Similarly, one would expect some differences between the design context of human artifacts and the design context of organisms. After all, the former are made intentionally and the latter are not. Organic design seems to be a lot more ad hoc, as though cobbled together with string and sealing wax, and not always very efficient or intelligent when compared with human design.

George Williams's example of the male reproductive system in humans makes this point precisely. And before him, of course, Darwin himself stressed the improvised nature of organic design: "Throughout nature almost every part of each living being has probably served, in a slightly modified condition, for diverse purposes, and has acted in the living machinery of many ancient and distinct specific forms" (Darwin 1862, 348)

Goal-Directed Systems Again

Grant that the reason we think it appropriate to use functional language in evolution is the metaphor—the Paley/Darwin metaphor—of design. Organisms, produced by natural selection, have adaptations, and these give the appearance of being designed. This is not a chance thing or a miracle. If organisms did not seem to be designed, they would not work and hence would not survive and reproduce. But organisms do work, they do seem to be designed, and hence the design metaphor, with all the values and forward-looking, causal perspective it entails, seems appropriate.

Now, let us think back to goal-directed or directively organized systems. I would say that humans are the apotheosis of goal-directed systems and that this points to two interrelated notions of teleology in the human world. On the one hand, we have human intentions and goals—the things that we want for ourselves and others, and also the thoughts and actions that we have and make and take to achieve these ends and goals. On the other hand, we have the objects—the artifacts—that we design and make to realize those intentions. In a way, therefore, an artifact's purpose is conferred on it by humans in order to achieve their own ends. The knife itself hardly has the end of cutting. *We* have the end of cutting, and so we design and make the knife.

I am not sure that our personal desires and intentionality require that we exhibit flexibility, the ability to regroup in the face of problems and adversity, persistence toward our ends in a range of ways—in short, that we show directively organized behavior. But surely directive organization is something intimately involved with human intentionality. To go back to the example at the beginning to this book: I want to get a Ph.D. so I enroll in a graduate program. As part of my course of study, I have to pass a language examination, and I fail the examination when I sit for it in French. Hence, I try again, but this time I take the examination in German. Now whether you approve of my actions, or whether you think I am really good enough to get a job when I am finished, at least you can understand what I am about and why I am about it. And more than this, you understand my actions to be intentional precisely because I keep at the task, trying to achieve my goal despite setbacks. No funny causation presents a problem here, but reference to the future is necessary to explain my behavior, and obviously values are intimately involved.

As I have said, I think we have two kinds of purpose here—personal goals and artifacts. In some respects, intentionality of the kind humans exhibit occurs in the animal world. Frans De Waal's studies of caged chimpanzees and Jane Goodall's studies of chimps in the wild contain example after example of all sorts of elaborate, end-directed behavior. When males are obstructed from their goals by competition from other males, they enlist the help of females in alliances and the like. And since the females have their own goals, getting their help can require a great deal of manipulation and persuasion and sheer hard work. I see no reason why one should deny the label of "intentionality" to such cases as these. Moreover, I think one can speak of the manufacture or use of "artifacts" by these brutes—sticks for poking or reaching and the like. They are simple and crude compared with human artifacts, but they are artifacts nonetheless.

The interesting and more controversial case comes as we move into the more metaphorical domain. How are we to regard nonthinking examples of directive organization? A torpedo is an artifact. Do we want to say that a torpedo with a homing device has an end-directedness over and above that of being an artifact with a function? And what about nonthinking, goal-directed systems in nature? What about mechanisms for achieving homeostasis, like sweating and shivering to keep the body's temperature constant? I confess that I do not have a definitive answer—nor am I sure that a definitive answer can be had. Grant (as we have granted) that we are probably never going to get adaptability unless we already have adaptation, do we want to say that nonthinking adaptability entails an end-directed dimension that is not already there with the adaptation? I suspect that some might think it does and others might think it does not. But that kind of open-endedness is precisely what metaphorical thinking allows. Some thinkers extend their metaphors more than others would. Ultimately, there is no absolutely right or wrong answer. The question is simply whether you find it particularly helpful to think in this way or not.

Wrong Focus

Once you have the metaphor of design in play, then of course you can ask questions about borderline instances and extensions and so forth. The real question, though, is whether, in the first place, the metaphor itself is an appropriate one. The question is not whether metaphors should ever be used at all but whether the specific metaphor of design should be used to explain evolution.

Darwinians argue strenuously that it must be used. Richard Dawkins speaks to precisely this issue, asking what job we expect an evolutionary theory to perform. Allowing that some people get excited about issues (such as the nature of species) that leave him cold, Dawkins agrees with John Maynard Smith that the "main task of any theory of evolution is to explain adaptive complexity, i.e. to explain the same set of facts which Paley used as evidence of a Creator." Jokingly he refers to himself as a "neo-Paleyist," concurring with the natural theologian "that adaptive complexity demands a very special kind of explanation: either a Designer as Paley taught, or something such as natural selection that does the job of a designer. Indeed, adaptive complexity is probably the best diagnostic of the presence of life itself" (Dawkins 1983, 404).

Formalists would beg to differ. They would raise the objection that adaptation and the associated design metaphor have their roots in British natural theology of the pre-Darwinian era, and they would ask again why, in the secular world of the twenty-first century, we should be bound by this retro-thinking. At the very least, they would like to see the metaphor used a lot less frequently. This worry has been voiced explicitly by the philosopher Peter Godfrey-Smith, who wondered if our interest in design in nature might not be little more than a reflection of something about our psychology. He distinguished among three concepts: empirical adaptationism, the belief that everything is a result of selection and has adaptive value; explanatory adaptationism, the view that adaptationism is the really big question facing evolutionists; and methodological adaptationism, the belief that one should look for adaptations when one is faced with organisms, especially new and unfamiliar ones. The first and third concepts are relatively unproblematic. No one holds the first in a strict form, and few deny the third. It is the second, explanatory adaptationism, that raises interesting and difficult issues. Godfrey-Smith (1999, 188) asks:

> Why should we infer that the phenomena that we happen to find puzzling are "the most important" phenomena? For an explanatory adaptationist like Dawkins or [philosopher Daniel] Dennett, the human eye is in the class of uniquely important phenomena for biology. The human toenail, probably, is not in that class. But toenails are just as real as eyes, and toenails have an evolutionary history as well. Rhapsodies about eyes might have a place in popular books and in public relations exercises, where rhapsodies about toenails might be less effective. But should the dispassionate biologist pay any attention to this? The "special status" of apparent design appears, in this light, to be analogous to the special status of cute and appealing animals in the conservation movement. Pictures of endangered koalas are more likely to elicit big donations than pictures of endangered spiders, but this fact has little scientific importance.

Godfrey-Smith is convinced that adaptationism is "nothing more than the personal preference of some biologists and philosophers; they find selection important because it answers questions that they happen to find interesting." He is of the opinion that, culturally speaking, the Darwinian approach to adaptation is crucially important inasmuch as it destroys the traditional theological argument from design and thus leads to a revised, naturalistic view of humankind. But this said, he warns that "it is also possible to place natural selection so much at the core of our view of our place in nature that there is massive distortion of that view" (Godfrey-Smith 2001, 351). He does not think the design perspective "vanishes altogether if we look at the world through wholly secular spectacles," but he does think "the design problem has been oversold in recent discussions" (p. 191). Ultimately, the point seems to be that we no longer have any good scientific reason for focusing on adaptation. "The roots of explanatory adaptationist thinking are found not so much in biological data as in views about the place of biology within science and culture as a whole."

Both sides of this debate about the design metaphor—speaking now specifically about explanatory adaptationism—seem to agree that an element of taste is at work here. An evolutionary biologist does not have to spend his or her time focused exclusively—or even chiefly—on adaptation. Other problems, like working out life's histories and the nature of variation and the reasons for speciation and more, compel. But this is true of any science. Some physicists like to work on galaxies, others on quarks. Yet, just as in physics, some issues in biology are fundamental, in that their solutions seem to have ramifications for much more. Working out the theory of forces has implications for so many fields in physics, and in a similar vein one might say that adaptation is just such an issue in biology.

The critic might respond that one has here a circular situation: Darwinians make searching for adaptation central to their program, and then when they find the adaptations they so fervently seek, they use them as support for Darwinism. But a better term than "circularity" might be "self-reinforcement." Darwinism is a successful theory—our scientific examples show that—and at the moment (and for the foreseeable future, whatever the qualifications) it is the only game in town, on its own merits. Fruit flies, dunnocks, dinosaurs, fig wasps—this is a theory on a roll. It has earned the right to set the agenda.

Adaptation is not the only evolutionary problem one can work on. A lot of the best evolutionary biologists today are working on development, where the functional issues are not foremost. But adaptation lurks in the background of most if not all evolutionary issues. Take the question of variation. Richard Lewontin has pointed out a huge amount of naturally occurring variation at the molecular level, as we have seen. In itself, one might say that these findings have little or nothing to do with adaptation. But in another sense, they have everything to do with adaptation. The search for variation was sparked by Dobzhansky's question about where the necessary fuel for natural selection could come from. Lewontin's finding of massive variation immediately sparked questions about whether it is held in populations through the workings of natural selection or whether it is the result of genetic drift or other nonadaptation-producing forces. These are adaptation issues. In physics today, we cannot get away from force. In biology today, we cannot get away from adaptation.

All of this is quite apart from questions about the extent to which the adaptation of organisms is ubiquitous and whether other putative causal factors should be getting more attention than they do. We have certainly seen no arguments that adaptation is nonexistent or that it is not widespread. More important, we have seen no arguments that adaptation is not a, if not the, fundamental issue. Dawkins is surely right that life is complex and that it does need explaining. If you are a biologist, you have made a commitment to explaining the complexity of life. And this holds true whether you are a Darwinian or not. So perhaps, until things start to grind to a halt, until the font of questions and answers dries up, perhaps we can put on hold Godfrey-Smith's worry that, from a broader historical perspective, the design problem has been "oversold." We have good scientific reasons for taking adaptation as the fundamental issue.

Of course, Lewontin and his school do not much care for many of the findings of the adaptationists. But to say that we should not play the game at all, or that we should count all as equal, requires some persuasive arguments. Better than arguments would be examples. Let those who worry about explanatory adaptationism show their dunnocks and dinosaurs and fig wasps. When they demonstrate that they can do science which explains and predicts without invoking adaptation even implicitly, then we can start to take their position seriously.

A Return to Reduction

We come back to the worries that started this philosophical part of the discussion. If we appeal to or use the concept of purpose or end-directedness, are we not caught in some kind of unacceptable vitalism or finalism? Does all hope of a satisfying and legitimate reduction go out the window? Is the status of physics and chemistry forever barred to evolutionary biology? Certainly, the worst of these worries has long been laid to rest. At the ontological level—the level that raises questions of existence—nothing we have seen has given any hint of appeal to special forces or impetuses or the like. Looking at organisms as though they were designed does not mean that they have vital forces. Anything but. They were made by natural selection, in a good old-fashioned mechanical way.

But when we come to the theoretical level of reduction—the level that raises questions about the deducibility of biology from physics and chemistry—matters are rather different. Here is surely an impassible block. Our analysis of biological functionality does not imply that we have reverse causality. No one is saying that the future-focused trilobite eye causes the present lens. But the analysis does imply that we are thinking in reverse fashion about causes—understanding what is happening in terms of what we expect will happen. The lens is understood in terms of what we think it will do—because, causally, we have a repeating circular situation, where lenses have done in the past what we think this one will do in the future. Moreover, we have a value component, as we judge entities—the parts of organisms—in terms of the overall benefit or good of the organism. It is hard to see how any amount of deduction (which is what a theoretical reduction is all about) is going to eliminate or explain away these factors. We may perhaps talk about eliminating the metaphor of design, but we are not going to translate it into nondesign-type language or understanding. Biology brings a different perspective to things, and that is a fact.

Should we be looking, if not to the translation of teleology in biology into nonteleology, to its elimination altogether? I am not now thinking so much of those who dislike adaptationism and who would like to see end-directed thinking diminished by doing away with adaptationism but rather of those who are adaptationists and who nevertheless think talk of purpose or function should be done away with. I refer now to Darwinians who are ashamed of what they take to be the weaknesses of their science. Responding to the Gould and Lewontin charges about overdoing the adaptationist strategy, Michael Ghiselin (1983) would have us "reject such teleology altogether." He would have us change the very questions we ask. "Instead of asking, What is good? we ask, What has happened? The new question does everything we could expect the old one to do, and a lot more besides" (p. 363). Less drastically, we have those like Ernst Mayr who would like to hide teleology by calling it a different name, "teleonomy." Some—philosophers mainly—would argue that since the obsession with ends comes in through the metaphor of design, a more mature science will emerge which drops the metaphor altogether and with it talk of purpose. "When you actually start to do the science, the metaphors drop out and the statistics take over" (Fodor 1996, 20). And even the philosopher Morton Beckner, a student of Nagel, likewise seemed to think that talk of purpose and design can go. "Even though teleological language cannot be translated away, there is a sense in which teleological language is eliminable. The sense is this: given any single case of an activity which is describable in teleological language, that case is also describable in nonteleological language. By this I mean that every observable aspect of the activity which in fact serves as the basis for our application of teleological concepts can also be described by means of a conceptual apparatus that is not teleological in character" (Beckner 1969, 162).

Of course, Aristotle, Kant, Cuvier, and Whewell would disagree with all of this. They would argue that if one is to think about organisms, one must think in terms of purpose and design. I am sympathetic to this older viewpoint, although I am wary of philosophical claims insisting that ways of thought must necessarily be as they have been. Perhaps, if making artifacts comes with being rational, then it follows that, as a matter of fact, serious thought about organisms will always be focused on end-directedness. But necessary or not, biology as we know it today would be dreadfully impoverished without a perspective that asks "what for"—and this includes molecular biology as much as traditional evolutionary biology, for the genetic code is as much part of the design metaphor as is the trilobite lens.

Indeed, for all that he thought elimination might be possible, this point about impoverishment was appreciated by Beckner: "Suppose we are watching a tank in which there is a single anchovy, and that we introduce a barracuda into the tank. At first nothing happens; then, when it would be reasonable for us to suppose that the anchovy spots the barracuda, the anchovy behaves as follows: he turns sharply, swims quickly to the surface, leaps out of the water, reenters, and repeats the sequence." This description (call it B) is entirely nonteleological, which certainly says more but in another sense says very much less than the teleological statement (call it A): "On sighting the barracuda, the anchovy engaged in an escape reaction." There is no translation. As Beckner points out: "A says less than B since B offers details not mentioned in A. It says more, since calling the anchovy's behavior an 'escape reaction' implies that it serves the function of escape, whereas this is not implied by B" (Beckner 1969, 162–163).

And that really is the nub of things. If we want a biology that is not interested in the reason why the stegosaurus has such a funny display on its back, is not intrigued by the peculiar shape of the trilobite lens, does not care why some butterflies mimic other butterflies, is unconcerned about the spirals of the sunflower, then presumably something can be done. Some low-grade biological activities like classification, perhaps, can indeed be done better than usual if we refrain from asking why: as we have seen, one of the big complaints of taxonomists is that adaptation makes the job difficult and that assumptions about evolution only cloud the picture. I presume we could also do a certain amount of embryology and physiology and other sorts of biological activity without any questions about adaptation—the sorts of things for which we saw Lauder and Amundson promoting a capacity-type analysis of function. But it is all surely going to be very limited—a classic case of cutting off your scientific nose to spite your philosophical face. Dobzhansky's most famous claim was that nothing in biology makes sense except in the light of evolution—that is, Darwinian evolution. And for many evolutionary biologists active today, the search for causes is here to stay.

But, to return to the worry that has haunted evolutionists from Wallace to Mayr: Is not the very use of a metaphor a sign of weakness? Should not evolutionists be striving to avoid it? I am not sure why they should. Here I stand with Charles Darwin. Avoiding metaphor is not something that physicists and chemists feel called upon to do. Of course, the logical positivists would probably have had us avoid metaphor, but we need not be bound by their strange demands any more than we need be bound by Paleyian natural theology. Today, the consensus is that, far from being a sign of weakness, metaphors are an essential part of thought—including scientific thought. They are a prime force in causing people to think in new, imaginative, and fertile ways. They are a key ingredient in the heuristic side of thought, one of the most prized virtues of great science. Metaphors do not provide answers so much as they raise questions. Precisely because one thinks (metaphorically) of organisms as if they were designed, one has many, many questions thrown up to be answered. Why are there plates on the back of stegosaurus? Why is the trilobite lens such a strange shape? Why does one butterfly mimic another? Why are pinecones and sunflowers patterned as they are? And much, much more. Without the metaphor, the science would grind to a halt, if indeed it even got started. The plates are there; the lens is there; the mimics are there; the swirls are there. Brute statements, which probably we would not be led to make in the first place because there would be no reason to even notice. Who would care about the trilobite lens? Why bother to describe it rather than something else? Whatever the ultimate logic of the case, at the practical, real-life level, Kant got it right. Biologists "are, in fact, quite as unable to free themselves from this teleological principle as from that of general physical science. For just as the abandonment of the latter would leave them without any experience at all, so the abandonment of the former would leave them with no clue to assist their observation of a type of natural things that have once come to be thought under the conception of physical ends" (Kant 1790, 25).

In a recent issue of *Science,* the cover picture, complementing the article inside, addressed the somewhat esoteric question of where one should locate the nostril on the sauropod dinosaur Diplodocus. In language that critics would have us avoid, the cover line read: "Nostril position in dinosaurs and other vertebrates and its significance for nasal function." In trying to solve this problem, the skeleton is not much help, because the hole in the bone is so huge (see page 271). Tradition put the nostrils up closer to the eyes, so that the supposedly aquatic creatures could stay submerged for long periods of time, with only their eyes and nose in the air. Now, however, thanks to comparative studies with dinosaur relatives, almost all of whom have their nostril down their snouts, opinion has changed. The nostrils have been moved down the face, nearer the mouth. And again using an approach that critics deplore, the scientific author shows just how important for him is the whole design-metaphor way of thinking—how it pushes him to make discoveries and to confirm hypotheses (Witmer 2001, 851):

> Thus, if the fleshy nostril were located in the traditional caudal position [up the face], then most of the narial apparatus would be out of the

> main airstream and instead in a cul-de-sac, which, from a design standpoint, seems problematic. On the other hand, if the fleshy nostril were located rostrally or rostroventrally [down the face], then the narial apparatus would be fully within the air flow, allowing the apparatus to participate more effectively in, for example, forced convective heat loss, facilitated evaporative cooling, and intermittent countercurrent heat exchange—processes that play a role in heat and water balance and in selective brain temperature regulation.

Purpose in evolution is obviously alive and well and mixing in the best circles! But what about the value component in the design metaphor? One does not have to be a logical positivist to insist that science ought to be value-free, as opposed to (say) religion and politics. Hence, inasmuch as the design metaphor brings in values—and this has been a significant conclusion of our analysis—we might fear that something has to be wrong with it. Unless, of course, we agree with those people today who would deny the very possibility of anything being value-free, including—especially including—science. People like Lewontin, who would argue that part of the oppressive power of science is that it pretends to be value-free when truly it represents all of the values of Western capitalist, patriarchal, racist society.

Fortunately, we do not need to open this particular can of worms here; we could point out that even the logical positivists agreed that science can be shaped by pragmatic values, aesthetic values, and the like. I have argued that methodological adaptationism is not merely a matter of taste, but taste certainly comes into the equation. Darwinism does show an interest in organized complexity, pushing it to the fore. Just as formalism shows its interests. Remember Thomas Henry Huxley (1900, 1.7–8): "What I cared for was the architectural and engineering part of the business, the working out the wonderful unity of plan in the thousands and thousands of diverse living constructions." So there is certainly a value component at this level, in Darwinism as in all sciences.

But what about the values inherent in the metaphor? What about the eye being "good" for humans and the leaf being "good" for the maple tree? This troubles Peter McLaughlin, a philosopher who manages to put down God, metaphor, and value all in one sentence. "As it stands, the assertion that natural functions are either metaphorical or divine is simply one particular variant of an antinaturalistic credo and is prima facie no less metaphysical than a commitment to intrinsic value in nature" (2001, 5). Perhaps so, although one thing we can point out—as we have already pointed out—is that the values being brought in by our metaphor are relative values rather than absolute values. A relative value is something on a scale of value that could be arbitrary, not endorsed by all. An absolute value is something on a scale of value that (supposedly) all accept as the true defining scale. In real life the distinction between absolute and relative values blurs. The Nazi would claim that Aryans are better than Jews on some absolute scale, a scale that the rest of us would deny. So perhaps it is better to say that an absolute value is something that the holder believes is universally acceptable and definitive.

However we make the cut, the value component introduced by the metaphor of design is relative—relative to an organism's well-being. What is the function of the nostril at the front of the face? To ensure good air flow, so that the brain can be heat-regulated in an efficient manner. What is the function of the trilobite's strangely shaped lens? To focus sharply in the aqueous environment in which it finds itself. Those trilobites without such lenses, or with such lenses only working partially, were at a survival and reproductive disadvantage. Science does not ask the question whether the trilobite's survival and reproduction is an ultimate or absolute good. Parasites such as viruses have parts that function well, only too well from the standpoint of some hosts, but no scientist says that a parasite is an ultimate or absolute good.

Absolute values about what is "good" are the problematic ones, the ones that we want to keep out of science. Relative values are commonplace. The difference is reflected in the distinction between "valuation" and "evaluation," the latter a phenomenon within science that even as conservative a philosopher as Ernest Nagel acknowledges and respects. Take the notion of "anemia." This refers to an animal having insufficient red blood corpuscles, which means that among other things it cannot keep the same constant internal temperature as normal animals in the species. But what is "normal" in this context? We have to make some kind of judgment call. An investigator or a physician will have to decide if the condition makes much difference to the animal's well-being. Can it function as well as other animals? If the animal is human, is he or she as comfortable as other people? We all have our ups and downs, but is this person outside the acceptable bounds? And so forth. The point is that some kind of valuing is going on here. "When the investigator reaches a conclusion, he can therefore be said to be making a 'value judgment,' in the sense that he has in mind some standardized type of physiological condition designated as 'anemia' and that he assesses what he knows about his specimen with the measure provided by this assumed standard" (Nagel 1961, 492). The key thing to note is that this is not an offensive kind of valuation—one is judging against a standard. One is making an "evaluation." And this is precisely what is happening in evolutionary biology. One is not making an absolute valuation. One is making a judgment against the standard of success in survival and reproduction.

So can one truly say that the trilobite lens is of value to the trilobite—even of value in a relative sense—when the trilobite almost certainly did not have full consciousness and an awareness of its needs? The serrated knife for cutting my bread is of value to me because—and precisely because—I have interests. That is where we came in. The trilobite has no interests. It just is. But that is surely the point at issue. *We* are seeing the trilobite as if it had interests. As if certain ends were of value to it. And that leads to understanding—*our* understanding. Whether or not we want to buy into all that he claimed about necessity, Kant was right in seeing that *we* do the science, and *we* try to make sense of the trilobite and its lenses. And we are doing this through metaphor—looking at the trilobite as if it had intentions and interests. As if it had values. The interests and values are not there in reality, but they are part of the way in which we map reality in order to make sense of it.

This is an uncomfortable conclusion for those who think the aim of science is to give an unvarnished report on reality. Rather like health foods—nothing added, nothing taken. Raw carrots and unpasteurized milk. But it is a realistic conclusion for those who think the aim of science is to make sense of reality. Like French cooking—squeeze and mold, add and reduce, cook and cool, until the dish is complete. Coq au vin and a bottle of Beaujolais.

Darwinian evolutionary biology is different. Because of the nature of organisms—their distinctive complexity—the design metaphor is appropriate and highly fruitful. The argument to organized complexity is precisely that—an argument to organization, to design (taken as a metaphor). This gives rise to a forward-looking kind of understanding—understanding in terms of final causes. A powerful kind of understanding that raises interesting and fruitful questions and stimulates important answers. It should be celebrated, not regretted.

CHAPTER 14

NATURAL THEOLOGY EVOLVES

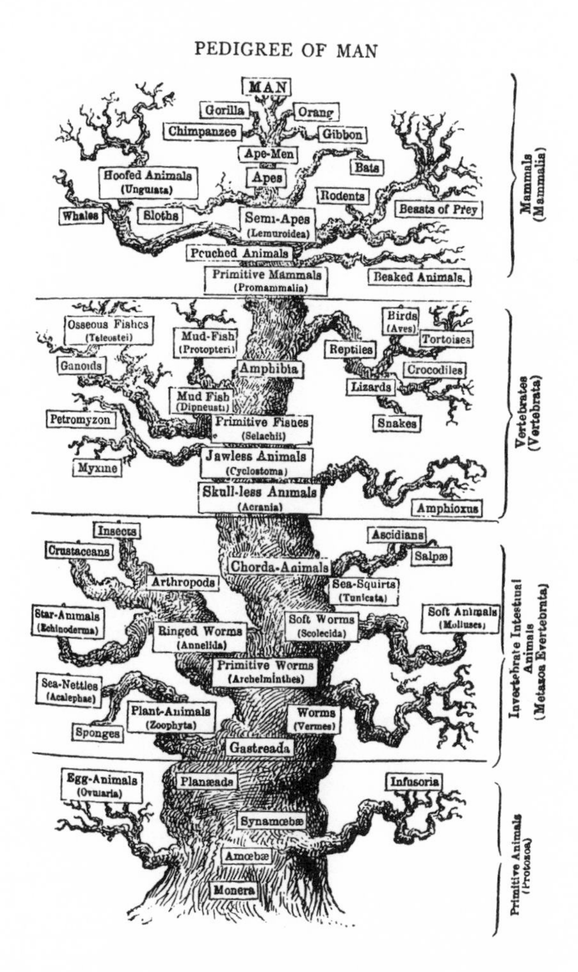

THE SPOTLIGHT TURNS now from the argument to organized complexity to the argument to design, taken in the literal, theological sense—to the argument for a creative intelligence. In the light of modern science, in the light of modern Darwinism, what can be said about inference to the existence of a deity? Does biology give us any understanding here, negative or positive? To set the scene, I begin with some historical background on Western Christianity. Darwinism is its child, after all, and it is within the family unit that relationships are most intense, fraught, and significant.

In most potted histories written about science, the Darwinian revolution looms as a mighty battle between science and religion. It is second only to the conflict two hundred years before, when the Catholic Church forced the aged Galileo to his knees, to recant his belief in the Copernican system of the heavens. After the *Origin,* bishops clashed with professors, politicians joked about descent, and radicals (not the least Karl Marx and his chum Friedrich Engels) celebrated the triumph of materialism. As is often the case, this common knowledge is a mishmash of truths, half-truths, and outright falsities. Bishops did indeed clash with professors: most famously, Samuel Wilberforce, Bishop of Oxford, and Thomas Henry Huxley, leader of the Darwinian support group, debated evolution in 1860 at the annual meeting of the British Association for the Advancement of Science. And politicians joked about descent: Benjamin Disraeli, leader of the Conservative Party and Queen Victoria's favorite prime minister, claimed to be on the side of the angels against the side of the apes. As for the radicals, Karl Marx considered dedicating part of *Das Kapital* to Darwin.

But on the other side, many Christians slipped very quickly and readily into acceptance of evolution. Before the ink was dry on the *Origin,* that theological enthusiast for unbroken law, the Reverend Baden Powell, was telling the world about "Mr. Darwin's masterly volume on *The Origin of Species* by the law of 'natural selection'—which now substantiates on undeniable grounds the very principle so long denounced by the first naturalists—*the origination of new species by natural causes:* a work which must soon bring about an entire revolution of opinion in favour of the grand principle of the self-evolving powers of nature" (Powell 1860, 139). The public that Disraeli and other politicians represented was often thrilled and somewhat shocked by evolution but quickly went over to a fairly general acceptance of the basic ideas. Chambers's *Vestiges* had prepared the way, and when someone with Darwin's authority endorsed evolution, many felt now that they could acknowledge their assent publicly.

Social theorists of all kinds took up evolutionary ideas, often very strange versions of Darwinism. For all his enthusiasm, Marx made it very clear that he wanted little to do with Darwin's industrialist metaphors and sought rather a higher (that is, Germanic) understanding of organic change. For every one who followed Darwin, ten more followed Herbert Spencer and his eclectic mix of middle-class British socioeconomic nostrums and quasi-continental morphological idealism.

A major reason why the story persists of the ongoing fervent opposition to evolution—an opposition led and backed by Christians of all kinds—is that it suited the Darwinians to tell such a story. Huxley and his friends were striving honorably and successfully to reform Victorian Britain, and they needed a kind of secular religion to oppose the forces of reaction, as represented by the Anglican Church. Tales like the row at the British Association, when supposedly Huxley crushed the Bishop, took on mythic status—somewhat like Moses's clash with Pharaoh—and grew in significance and triumph with the telling and retelling. Like the robbers who set about Falstaff, the defeated opposition grew without limit. However, modern scholarship has shown that, at the event itself, the Bishop and his pals felt well satisfied with their efforts, and everyone probably went happily off afterward to the same shared supper. No one got boils. No first-born sons were slain.

Not, let me hasten to add, that there was no theological opposition to

Darwinism in particular and evolution in general. Disraeli may have joked about evolution—he joked about many things—but his great opponent, William Gladstone, fervent member of the Church of England, went to his grave convinced that Genesis is our best authority on origins. In later life, he and Huxley had a series of exchanges, one in particular over the significance of the Gaderene swine, into which Jesus transferred devils, at which point the swine jumped off a cliff. The two old men had a wonderful time discussing whether this was an unwonted abuse of other people's property.

While firm generalizations about these matters are very difficult to make, around 1870 or a little later in Britain those with religious inclinations started to follow the rest of the reading and thinking public into acceptance of some form of evolution. Given that Darwinism was homegrown and that English Protestantism contained a high proportion of Anglicans, who never had as extreme a Bible-centered faith as nonconformists like the Methodists and Baptists, I am not sure how strident and strong the early opposition was. Adam Sedgwick, Darwin's old geology teacher, fulminated and spluttered, but he fulminated and spluttered about everything and, besides, he was a force of the 1830s, not the 1860s. Others, like the Duke of Argyll, head of the Campbell clan, patrician politician, father-in-law of one of Victoria's offspring, and general man of science, had little time for unaided natural selection but granted (in an Asa Gray-type fashion) a sort of guided evolution. "Creation by Law—Evolution by Law—Development by Law, or, as including all those kindred ideas, the Reign of Law, is nothing but the reign of Creative Force directed by Creative Knowledge, worked under the control of Creative Power, and in fulfilment of Creative Purpose" (1867, 293–294).

In America, opposition to evolution probably continued a little longer than it did in Britain, being fairly general until about 1875. Obviously it was not universal—Asa Gray was an evolutionist as well as a dedicated evangelical. After 1875, in America as in Britain, the realization grew among the religious that evolution was a theory here to stay, one now accepted by the scientific community. Hence, Protestants (theologians, preachers, then laypeople) moved toward some kind of acceptance of evolution. By about 1900, among Protestants in the English-speaking world, about 75 percent accepted evolutionism in one form or another

—usually a form that allowed for direction to humankind and downplayed pure selectionism. Seventh Day Adventists and Mormons universally rejected even this kind of evolution; Unitarians universally accepted evolution and may have given selection some role; and others fell in between. Geographical location was significant, with generally more acceptance of evolutionary ideas in the American North than in the South. This pattern may have been reversed in Britain, but in any case many exceptions were found in both countries.

After 1900, although these proportions probably do not change greatly, the situation becomes rather more complex. Those who rejected evolution of any kind became much more vocal, especially in the United States, and by the 1920s, fundamentalists sparked the famous "monkey" trial in Dayton, Tennessee, when a young schoolteacher, John Thomas Scopes, was prosecuted for teaching evolutionary ideas. The rise of anti-evolution sentiment in America was occasioned less by theological and scientific than by social issues—the success of the anti-liquor movement and its quest for new targets, the disruptions of World War I and the popular view that German militarism was a function of Social Darwinism, and perhaps most important the rise of public education, which exposed children to ideas that their conservative parents and pastors saw as not just false but as an ideology in its own right, hatched up by a group of Yankee intellectuals in cities like Boston, New York, and Chicago. But without saying that this whole movement was ephemeral—it was not, although it does seem to have peaked and then faded somewhat—the fact is that most Protestant Christians in America, as well as in the British Empire, had made their peace with evolution in some form or another and were not about to return to the worldview of the early part of the nineteenth century. And so it continues today, especially in the United States, where religion plays a much greater role in American culture than it has for many years in Britain.

To a great extent, Roman Catholics sat out the Darwinian revolution. Besides having had their fingers burned over the Galileo episode, they had other more pressing fish to fry. In both Britain and the United States, Catholics tended to be of the lower and immigrant classes—Irish in England and Irish and European (especially Mediterranean and Eastern European) in America. They were fighting for survival, for status, for a future for their children, and they had little time for esoteric debates about evolution, especially given that their Church had never had the intense biblical focus of the Protestants. But there was some interest in the topic, and for the most part it followed the Protestant pattern. Some Catholics accepted evolution of a directed form and insisted that it could be reconciled with the teachings of the Church, provided that space was left for the miraculous creation of human souls.

This was the position of John Henry Newman. We have seen already that he never had much taste for natural theology, so it was never a barrier to acceptance of evolution. The same is true of the literal words of the Bible. "The Fathers are not unanimous in their interpretation of the 1st chapter of Genesis. A commentator then does not impute untruth or error to Scripture, though he denies the fact of creation or formation of the world in six days, or in six periods. He has the right to say that the chapter is a symbolical representation, for so St. Augustine seems to consider" (letter of 1864, in Newman 1971, 266). More positively, Newman always had a somewhat sympathetic attitude toward science. As an undergraduate he had attended Buckland's geology lectures and liked them quite a lot, although he was not keen on attempts to tie in geology with Genesis. Then, when he was working to set up a Catholic university in Dublin in the 1850s, Newman spoke strongly of the need to keep scientific inquiry unfettered by religious dogma.

Newman's own position was that science and religion deal with different spheres and thus, properly understood, do not interact and cannot conflict. Speaking of the natural and supernatural worlds, he said that "it will be found that, on the whole, the two worlds and the two kinds of knowledge respectively are separated off from each other; and that, therefore, as being separate, they cannot on the whole contradict each other" (Newman 1873, 389). Speaking explicitly in favor of Bacon, Newman spoke positively of a "philosopher" who put himself against scientists and theologians who were "occupied with the discussion of final causes, and solving difficulties in material nature by means of them" (p. 401).

Keen to defend dogma and the authority of continued revelation through the Church (as opposed to static biblical literalism), Newman always argued for an evolutionary attitude toward knowledge and belief. In his *Essay on the Development of Christian Doctrine,* the work that marked his move from Anglican to Roman Catholic, Newman quoted Bishop Butler with approval: "The whole natural world and government of it is a scheme or system; not a fixed, but a progressive one; a scheme in which the operation of various means takes up a great length of time before the ends they tend to can be obtained" (Newman 1845, 74, quoting Butler 1736, ii.4). After publication of the *Origin,* Newman made it clear that he found no problems at all with biological evolution per se. To the conservative Anglican Edward Pusey, in support of Darwin's receiving an honorary degree from Oxford University, Newman wrote on June 5, 1870 (1973, 137):

> 1. Is this [Darwin's theory] against the distinct teaching of the inspired text? If it is, then he advocates an antichristian theory. For myself, speaking under correction, I don't see that it does—contradict it.
>
> 2. Is it against Theism (putting Revelation aside)—I don't see how it can be. Else, the fact of a propagation from Adam is against Theism. If second causes are conceivable at all, an Almighty Agent being supposed, I don't see why the series should not last for millions of years as well as for thousands.

In fact, Newman was not enthused by the details of Darwin's theory. Falling in with general opinion, he thought that there had to be more to it all than unaided natural selection. And Newman's separation of religion and science left no room for blind chance in the arrival of humans or the creation of souls. But evolution in itself was no problem at all for him.

At the beginning of the twentieth century, paralleling the rise of fundamentalism (although for different reasons), the Church in Rome clamped down on those who would go their own ways and try to make some accommodation with the modern world. Evolution was one of those ideas that fell into bad odor. Father Pierre Teilhard de Chardin, a French Jesuit paleontologist and the most significant of thinkers who tried to work through a synthesis of evolution and Christianity, was not allowed to publish his works in his lifetime. (He died in 1955.) Only gradually, as the century progressed, was this kind of oppression lifted.

Pope John Paul II, although doctrinally very conservative, was always friendly toward science—after all, he was a faculty member of Krakow

University, whose first famous teacher was Nicholas Copernicus. With the exception of the soul, the Pope allowed evolutionary thinking full reign. More than this, he seemed to favor the dominant causal paradigm (John Paul II 1997, 382). "It is indeed remarkable that this theory has been progressively accepted by researchers, following a series of discoveries in various fields of knowledge. The convergence, neither sought nor fabricated, of the results of work that was conducted independently is in itself a significant argument in favor of this theory."

The Big Disjunction

To recap, after the *Origin,* religious thinkers (Protestants and Catholics) either had to accept a traditional, Paleyite form of the argument to design, which held that the only way one can explain the hand and the eye is through some kind of nonnatural intervention, or they had to accept some form of evolution, which is capable of producing contrivance according to natural laws. Dawkins has said that only after the *Origin* could one be an intellectually fulfilled atheist, and he has a point. Before Darwin, whatever Hume may have said, adaptation—the argument to complexity—seemed explicable only by some form of the argument to design. It was the best explanation. After Darwin, this simply was not so. Which leads to a dilemma. The choices were to reject the argument to design *in toto*—the position taken by Huxley with joyful confidence and by Darwin with discomfited hesitation—or to revamp the argument to design by making a virtue of evolution. Or—a third option—you can reject evolution, keep to the traditional argument to design, perhaps updated by modern science, and tough it out. In the time since the *Origin,* we have seen all of these options in action—a pattern, incidentally, that was anticipated by the scientists of the era, as we saw in earlier chapters.

Start with the period up to about 1875, when the evolutionary hypothesis was still regarded with much suspicion by religious believers. The traditional argument from design continued to rule and was taken as a direct counter to Darwinism, however one interpreted that particular theory. Sedgwick (1860, 103) included this point in his charges, complaining that his "deep aversion to the theory" was in part based on the fact that it "utterly repudiates final causes." More sophisticated was Professor Charles Hodge of Princeton Theological Seminary, a conservative Presbyterian and one of the leading theologians of his time. He stood four-square behind the traditional argument to design. It did not prove every aspect of the Christian God, but it did prove—really prove—his creative and concerned nature. This, and like arguments, "have been regarded as sound and conclusive by the wisest men, from Socrates to the present day" (Hodge 1872, 1.203). Final causes, evidences of design, exist, and they are definitive. "The doctrine of final causes in nature must stand or fall with the doctrine of a personal God. The one cannot be denied without denying the other. And the admission of the one involves the admission of the other. By final cause is not meant a mere tendency, or the end to which events either actually or apparently tend; but the end contemplated in the use of means adapted to attain it" (p. 227).

Not that Hodge was content to rest simply on stark statements of natural theology. He gave lots of examples—the Bridgewater Treatises were explicitly acknowledged resources—and much was familiar. To no one's surprise, the eye had a starring role, for it "is the most perfect optical instrument constructed in accordance with the hidden laws of light." And indeed, "If the eye, therefore, does not indicate the intelligent adaptation of means to an end, no such adaptation can be found in any work of human ingenuity" (p. 218). Hodge trotted out all of the scientific objections to Darwinism. The failure of artificial selection to produce new species, for example. But the conclusion was clear and with unmistakable consequences for our thinking about design. "Ordinary men reject this Darwinian theory with indignation as well as with decision, not only because it calls upon them to accept the possible as demonstrably true, but because it ascribes to blind, unintelligent causes the wonders of purpose and design which the world everywhere exhibits; and because it effectually banishes God from his works" (p. 30).

After about 1875, wonderful fossil discoveries from the American West started to pour in, including the fabulous sequence demonstrating the transition of the horse from a tiny animal running around on all five toes to the modern animal that is much larger and has but a single toe or hoof. Among professional evolutionists, the fossil record has always been but one part of the story, and not necessarily the most important or most convincing part. But there is nothing like a handful of bones for persuading the general public. Now, as the religiously inclined were pulled on board—and this was not always with reluctance, especially among Calvinists, who put a premium on the great significance of scientific understanding—we start to see a shift in attitudes toward the argument to design. At this point the disjunction kicked in. No longer could one simply adhere to the Paley-type position that adaptation implies design implies God. The lawbound creation of species means that God is pushed aside in this respect. So now the challenge was to reintroduce final cause in a somewhat different mode, and the problem was solved in two complementary ways.

First, the absolute ties between design and Designer were loosened. The move was from the position of a Hooker or Hodge—that organized complexity implies God, necessarily—to the position of an Aquinas or Whewell—that organized complexity fills out our understanding of God already accepted on grounds of faith. Second, a virtue was made of necessity, and in line with the general progressivist bent that secular theories of evolution were now adopting, the progress culminating in humans was taken as a sign that all was not random but planned by the Great Evolutionist. Theological concerns about the conflict between progress and Providence were downplayed or forgotten, despite an undercurrent of worry that progress might not be inevitable and that degeneration and collapse might be our actual fate.

In other words, a divinely driven, historical teleology started to impinge on the story of evolution in a major way. The move was made from a God-inspired, short-term, adaptive final cause to a God-inspired, long-term, historical final cause: from the planned ends of organisms to the planned end of creation. In England, Frederick Temple, future Archbishop of Canterbury, writing in the 1880s, was explicit on the need to make this shift from an old-style natural theology to a new-style evolution-informed substitute. He stressed that, after Darwin, if one were rewriting the *Natural Theology* of Paley, the emphasis would be different. "Instead of insisting wholly or mainly on the wonderful adaptation of means to ends in the structure of living animals and plants, we should look rather to the original properties impressed on matter from the beginning, and on the beneficent consequences that have flowed from those properties" (Temple 1884, 118–119).

Across the Atlantic and back at Princeton, the new president, James McCosh, was a Presbyterian divine who started out in Edinburgh, moved to Northern Ireland, and then cast his lot with the New World. Sensitive to science and yet deeply committed to his faith, he too saw evolution as the key to a new natural theology rather than the slayer of the old. He invited us to "look upon evolution as we do upon gravitation as a beneficent ordinance of God. Gravitation is a law of contemporaneous nature, binding the bodies in space. Evolution is a law of successive nature, binding events in time. The two are powerful instruments in giving a unity and consistency to the world, and in making it a system compacted and harmonious, admired by the contemplative intellect" (McCosh 1887, 1.217). He spelled out in detail the connection between progress, evolution, and design. "There is proof of Plan in the Organic Unity and Growth of the World. As there is evidence of purpose, not only in every organ of the plant, but in the whole plant . . . so there are proofs of design, not merely in the individual plant and individual animal, but in the whole structure of the Cosmos and in the manner in which it makes progress from age to age." But do not think that any of this happens by chance. God had to arrange and drive everything, including—especially including—those things covered by law (McCosh 1871, 90–92).

So much for Protestants. Catholics writing on the subject were few, and essentially their position—until the Church clamped down on everything—was little different. St. George Mivart's *Genesis of Species,* first published in 1871, tore into natural selection with vigor, arguing that pure Darwinism is riddled with problems. In its stead, Mivart—who never once doubted the fact of evolution—supposed (as did Huxley) that sometimes major jumps or leaps, saltations, occur, and he supposed (as Huxley did not) that these were directed and produced adaptations—adaptations that are the design products of a good God. Not that Mivart was insensitive to the apparent flaws in design and the places where things simply seem to go wrong. "Organic nature then speaks clearly to many minds of the action of an intelligence, resulting on the whole and in the main in order, harmony, and beauty, and yet of an intelligence the ways of which are not as our ways" (p. 273). But overall, the picture is one of explicit design, if now somewhat attenuated. "Teleology concerns the ends for which organisms were designed. The recognition, therefore, that their formation took place by an evolution not fortuitous, in no way invalidates the acknowledgment of their final causes if on other grounds there are reasons for believing that such final causes exist" (p. 277).

Souls of course are created miraculously. But evolution is responsible for the rest, culminating in our species, *Homo sapiens.* And let there be no mistake, evolution or not, we are into something altogether new and better. "With our entrance on the world of self-consciousness, reason and free volition, we impinge upon another order of being from that revealed to us by all below it—an order of being which the cosmical universe, as it were, intersects, as the different lines of cleavage and stratification may intersect in the same rock" (Mivart 1876, 198–199).

The Twentieth Century

Fundamentalism centered on a strong drive by evangelical Protestants to return to a strict, literally read, biblically based faith. Almost paradoxically, the pamphlets that gave the movement its name—*The Fundamentals,* published in the second decade of the century "Compliments of Two Christian Laymen" and "sent to every pastor, evangelist, missionary, theological professor, theological student, Sunday school superintendent, Y.M.C.A. and Y.W.C.A. secretary in the English speaking world, so far as the addresses of all these can be obtained"—contained some essays that were sympathetic to a guided form of transmutationism. But generally fundamentalists strongly opposed evolution, albeit not entirely on natural theological grounds.

Consider the material surrounding the Scopes trial, particularly the writings of William Jennings Bryan, three-times presidential candidate and brilliant orator, who led the team for the prosecution. Undoubtedly, threats to the argument to design were factors. "It would seem that a knowledge of nature would be sufficient to convince any unprejudiced mind that there is a designer back of the design, a Creator back of the creation, but . . . some of the scientists have become materialistic" (Bryan 1922, 24). But much more offensive was the blunt contradiction between evolution and the early chapters of Genesis taken literally. And certainly for Bryan, who had served for a while in Woodrow Wilson's cabinet before resigning because of what he (rightly) saw as a move toward militarism and involvement in World War I, the perceived connection between evolution and Social Darwinism was at least as significant, if not more so, than any troubles for natural theology.

Among more mainstream theologians, the thinking remained much as it had been in the last quarter of the nineteenth century. The Gifford Lectures of Frederick Tennant, given in the 1920s, are quite explicit in their belief that what counts from an end-directed perspective—what is really important to the Christian who turns to science for enlightenment and understanding—is not the question of organic adaptation. It is rather the question of progress, of human status and destiny. These are the real issues with which one must grapple.

And then there was Teilhard de Chardin. Proudly and unambiguously, he put progress at the heart of his vision. Teilhard (1959) saw life evolving upward through the biosphere, to the realm of humans and consciousness (the noosphere), and then even further onward and upward to the Omega Point, which in some way he identified with the Godhead and with Jesus Christ. Greatly influenced by people like Henri Bergson, Teilhard obsessed not about adaptation but about progress: "An ever-ascending curve, the points of transformation of which are never repeated; a constantly rising tide below the rhythmic tides of the ages—it is on this essential curve, it is in relation to this advancing level of the waters, that the phenomenon of life, as I see things, must be situated" (p. 101). And so on and so forth.

Bergson also influenced one of the most articulate Protestant voices in the mid-century, Canon Charles Raven of Cambridge University, author of works on John Ray and ardent reconciler of science and religion. By now, the scientific community was fully into the revival of Mendelian-infused Darwinism, but for Raven it was of little account. Genetics, if anything, just makes things worse, because the randomness of mutation renders even more difficult any proper account of nature—an account that can speak adequately to God's design and purpose. Somehow, we need a kind of reinvigorated Lamarckism, one that will be consonant with modern science and yet give direction to the unfurling of life. "Although the causes of mutations are almost wholly unknown, the Darwinian theory that they arise fortuitously is not consistent with the sudden appearance of new types of structure and behaviour involving the simultaneous accomplishment of a number of separate but interdependent changes . . . Here a principle regulating their co-ordination, a principle which displays the evidence of initiative and of design, seems manifestly at work" (Raven 1943, 107). In his Gifford Lectures, giving his own position on the evolutionary process, Raven (1953, 146) wrote:

> On the biological side the evidence for continuity, creativity and design, if taken in connexion with the data already examined from the religious experience of mankind, is compatible with an interpretation of the Godhead which Christians will recognize. It suggests the theory that there is in the whole evolutionary process a purposive urge promoting not only larger ranges of activity, fuller individuation and ultimately the emergence of personality, but a harmony in diversity, the gradual fulfilment of a plan, the integration of the several elements of the design into a complex and inclusive pattern.

Toward the end of his life, Raven became an enthusiast for the ideas of Teilhard de Chardin.

The Neo-Orthodox Critique of Natural Theology

In the first half of the twentieth century a movement called neo-orthodoxy swept Protestant theological thought. Although the movement did not stem directly from the encounter between science and religion, powerful voices within it argued against the legitimacy of any kind of natural theology—including any attempts to support or resuscitate the argument to design. Harking back at least to the nineteenth-century Danish thinker Søren Kierkegaard, who had argued that genuine faith can only come in the face of radical uncertainty, the most significant thinker of the century, the Swiss theologian Karl Barth, urged a retreat from the optimistic God of progress, back to the stern God of Providence and grace.

Barth painted and promoted a picture of a deeply sinful humanity that can be saved by, and only by, God's freely given and unearned forgiveness, or grace. In the light of the horrors of World War I and the rise of the National Socialists in Germany, Barth argued that any attempt to reason one's way to God and his nature is fallacious—faith and faith alone will take us there. And the God to which we are taken by faith is the

Christian God—the one who gave his Son to die on the Cross for our sins—rather than merely to the distant God of the Greek philosophers. Barth thought that ultimate reality is in some sense quite other and different from experience as we normally know it. It is not just strange but totally at right angles. It is a question of different conceptual frameworks. We have our world, "the world of human beings, and of time, and of things . . ." And then we have the unknown world, "the world of the Father, and of the Primal Creation, and of the Redemption." These different dimensions are brought together for Barth through the person of Jesus Christ, our Savior: "The point on the line of intersection at which the relation becomes observable and observed" (Barth 1933, 29). Natural theology therefore is flawed through and through. It is a mistaken enterprise.

Expectedly, some people fought back, arguing that even on biblical grounds—a major plank for Barth's Christological focus—one can find justification for natural theology. "The heavens declare the glory of God: and the firmament sheweth his handy-work" (Psalm 19.1). Likewise, Saint Paul—an educated man, comfortable with Greek thought—used natural theological modes of discussion (Romans 1.20). Not even this line of argument was needed by those who used biological theories to bolster their faith. They simply argued that Barth was mistaken in his claims, heretically blind to the glory of nature. Raven became somewhat unglued on the subject, and criticisms of Barth were littered in his writings like Mr. Dick's references to King Charles's head.

But for many Christians, both Protestants and Catholics, the neo-orthodox critique of natural theology has had an effect. They recognize more fully the place, the strengths, and the limitations of any natural theological approach, including—especially including—one that makes appeal to design. The eminent philosopher Alvin Plantinga, writing from the Reformed tradition, allows a natural theology that might help nonbelievers to faith but prefers himself to go the route of Calvin, to see a belief in God as basic, planted in humans by God himself, who thus "revealed himself and daily discloses himself in the whole workmanship of the universe" (Plantinga 1983, 66, quoting Calvin's *Institutes,* 50).

In a like manner, the Lutheran theologian Wolfhart Pannenberg—a one-time somewhat uneasy student of Barth (a man who was not much given to criticism from the young)—argues that one can no longer subscribe to a traditional natural theology. In its stead he argues for what he calls a "theology of nature." This is the kind of shift we have seen from Paley to Whewell, where one abandons the hope of strict and conclusive proof and aims for illustration of and support for something that one has accepted already on other grounds. It is, of course, absolutely the position of Newman, who—Protestant or Catholic—was a more subtle theologian than any. Not that this stance is wishy-washy and second-rate. Pannenberg writes that: "If the god of the Bible is the creator of the universe, then it is not possible to understand fully or even appropriately the processes of nature without any reference to that God. If, on the contrary, nature can be appropriately understood without reference to the God of the Bible, then that God cannot be the creator of the universe, and consequently he cannot be truly God and be trusted as a source of moral teaching either" (Pannenberg 1993, 16). Science does its work and theology gives it meaning: "abstract knowledge of regularities should not claim full and exclusive competence regarding the explanation of nature; if it does so, the reality of God is denied by implication."

Putting matters in the language of values, the position is that one cannot simply read absolute values from nature—one cannot simply conclude that the world proves the interests and intentions of a loving god. But if one (on faith) approaches nature with the conviction that nature is of value (of supreme value), then this conviction is reinforced and illustrated by nature. A feedback loop exists between the believer and the world into which God has put him or her. Even Catholics today are not too far distant from this position, hardly a great surprise given that they have always stressed the priority of revealed religion over natural religion. Consider the recent encyclical letter, *Fides et Ratio* (1998), of Pope John Paul II. We begin with faith, with the revelation that Jesus is the Savior, the Messiah. This knowledge could not come through reason. But then we move to reason in the light of this. "Revelation has set within history a point of reference which cannot be ignored if the mystery of human life is to be known. Yet this knowledge refers back constantly to the mystery of God which the human mind cannot exhaust but can only receive and embrace in faith. Between these two poles, reason has its own specific field in which it can enquire and understand, restricted only by its finiteness before the infinite mystery of God" (section 14). In articulating the proper position of the Christian thinker, the Pope stresses the significance of faith and (citing Kierkegaard with approval) points to the limitations of reason. Indeed, even when reason operates, it can do so only in a limited fashion, with faith as its foundation.

John Paul unambiguously puts Saint Thomas on a pedestal, praising his contributions to our thinking. "Aquinas was keen to show the primacy of the wisdom which is the gift of the Holy Spirit and which opens the way to a knowledge of divine realities. His theology allows us to understand what is distinctive of wisdom in its close link with faith and knowledge of the divine" (section 49). Given the significance for Saint Thomas of the argument to design, and given the Pope's warmth toward modern evolutionary ideas, one concludes that for Pope John Paul II, acceptance of Darwinism does not threaten our teleological understanding of the world, a teleological understanding which leads to apprehension and appreciation of the creator.

Where Religion Stands Today

Few in the theological mainstream today would urge that natural theology can prove God absolutely. What does this mean for the argument to design? My sense is that the tradition of the past century and a quarter continues. The significance of the argument to organized complexity is deemphasized, and the progressive move up the ladder of evolution to our own species is played up. (This might be less a theological move and more a function of the fact that many of today's leading commentators on the science/religion relationship have a background in the physical rather than the biological sciences.) Combined with this progressivism is a distrust of natural selection's ability to deliver the goods. Given that Darwinism is now the accepted scientific paradigm and that people can no longer simply appeal to the authority of the best workers in the field, what we now find is more argument about the need for a new or revised evolutionism, one that guarantees the inevitable arrival of humankind on the planet.

Entirely typical is Holmes Rolston III, a philosopher of the environment, ordained Presbyterian, trained physicist, and recent Gifford Lecturer. He takes as his text a line from one of my books: "The secret of the organic world is evolution caused by natural selection working on small, undirected variation." And he makes clear at once that the problem is that of design and direction (Rolston 1987, 91):

> The troublesome words are these: "chance," "accident," "blind," "struggle," "violent," "ruthless." Darwin exclaimed that the process was "clumsy, wasteful, blundering, low and horribly cruel." None of these words has any intelligibility in it. They leave the world, and all the life rising out of it, absurd. The process is ungodly; it only simulates design. Aristotle found a balance of material, efficient, formal, and final causes; and this was congenial to monotheism. Newton's mechanistic nature pushed that theism toward deism. But now, after Darwin, nature is more of a jungle than a paradise, and this forbids any theism at all. True, evolutionary theory, being a science, only explains *how* things happened. But the character of this *how* seems to imply that there is no *why*. Darwin seems antitheological, not merely nontheological.

So what are we going to do instead? We have to have some way by which nature guarantees order and upward direction. Apparently we need (and have) a kind of rachet effect, whereby molecules spontaneously organize themselves into more complex configurations that then get incorporated into the living being and thus move the living up a notch, from which it seems there is no return. These are the sorts of events—"order for free"—promoted by Stuart Kauffman and others. Rolston writes: "But to have life assemble this way, there must be a sort of push-up, lock-up effect by which inorganic energy input, radiated over matter, can spontaneously happen to synthesize negentropic amino acid subunits, complex but partial protoprotein sequences, which would be degraded by entropy, except that by spiraling and folding they make themselves relatively resistant to degradation. They are metastable, locked uphill by a ratchet effect, so to speak, with such folded chains much upgraded over surrounding environmental entropic levels. Once elevated there, they enjoy a thermodynamic niche that conserves them, at least inside a felicitous microspherical environment" (pp. 111–112).

This yields a kind of upward progress to life. For all that every "form of life does not trend upslope," finally, as expected, we move up to humankind, God's special creation (pp. 116–117). At this point, Rolston (echoing many earlier Christian progressionists) ties in the upward struggle of evolution with the upward struggle of humankind:

> What theologians once termed an established order of creation is rather a natural order that dynamically creates, an order for creating. The order and newer accounts both concur that living creatures now exist where once they did not. But the manner of their coming into being has to be reassessed. The notion of a Newtonian Architect who from the outside designs his machines, borrowed by Paley for his Watchmaker God, has to be replaced (at least in biology, if not also in physics) by a continuous creation, a developmental struggle in self-education, where the creatures through "experience" becomes increasingly "expert" at life.

Do not be perturbed by this. "This increased autonomy, though it might first be thought uncaring, is not wholly unlike that self-finding that parents allow their children. It is a richer organic model of creation just because it is not architectural-mechanical. It accounts for the 'hit and miss' aspects of evolution. Like a psychotherapist, God sets the context for self-actualizing. God allows persons to be imperfect in their struggle toward fuller lives . . . , and there seems to be a biological analogue of this. It is a part of, not a flaw in, the creative process" (p. 131). We could be back in the age of Thomas Henry Huxley and Asa Gray. We could be reading James McCosh or William Temple—although one suspects that the latter, who, when headmaster of Rugby School earned a reputation as a disciplinarian ("a beast, but a just beast"), would have had God recommending caning, cricket, and cold baths rather than a session on the therapist's couch.

Darwinism and Progress

Even if one is moving from a natural theology to a theology of nature, one can surely see and empathize with the reasons that led religious people, in the age of evolution, to try to recover something from the ruins. A Paley-type argument to design may no longer work, but there are obvious attractions to a theology that includes a special status for humankind as a central part of the solution. The new argument to design—or (if not an argument) the new picture that sees life as progressing upward to our species—was a good move for the Christian against the background of the development of science. Nevertheless, empathy for the strategy taken does not in itself make it right. Although something may have been an understandable and defensible move in light of the social situation in the years after Darwin, it does not follow that it is an understandable and defensible move in the light of modern evolutionary biology.

Holmes Rolston promotes progressionism because (and precisely because) he is not a Darwinian. For him, self-organization can do what is needed. But what if you are a Darwinian and you think that self-organization has been dramatically oversold? What if you think that natural selection still rules triumphant? Does the whole attempt to save even some vestige of the argument to design—the new version, progress-based, human-ending—come crashing down? Some Darwinians do indeed see progress and believe that selection—particularly the selection produced by arms races—brings it about. Others are far less certain. Apparently, you can pay your money and take your choice. For those to whom these issues matter, resources can be found for the Darwinian who would argue that the arrival of humankind—somewhere, somehow—was not purely fortuitous. But whether they are enough to give one confidence in a revised, post-Paley theology of nature is another matter.

Must this then be the final word? Those Christians committed to natural theology were right to be concerned about Darwinism, and it seems that their concerns have not yet finally abated. Can a way forward be found that stays true to Darwinism, puts aside worries about progress, and yet satisfies the believer? One could simply go the way of Newman, declaring science and religion to be different spheres and hence noninteracting. Theological issues like the special status of humans are to be given theological solutions and not at all to be solved in terms of science, Darwinian or otherwise. Augustine argued that God stands outside time, and hence for him the thought of the creation, the act of the creation, and the completion of the creation are as one. Thus, the arrival of humankind was guaranteed from the first, and the specific mechanism of production is theologically irrelevant. But is the cure now worse than the disease? One has made Darwinism nonthreatening by making it irrelevant.

Is there a *via media* (to use the term that the Anglicans like to use of themselves) between irrelevance and contradiction? For the Christian, can Darwinism be more than irrelevant and less than contradictory? This must be our final question, although to get to it we must first tear down the barriers of those who—for starkly different reasons—argue that the road is closed and Darwinism and theology must be forever apart.

CHAPTER 15

TURNING BACK THE CLOCK

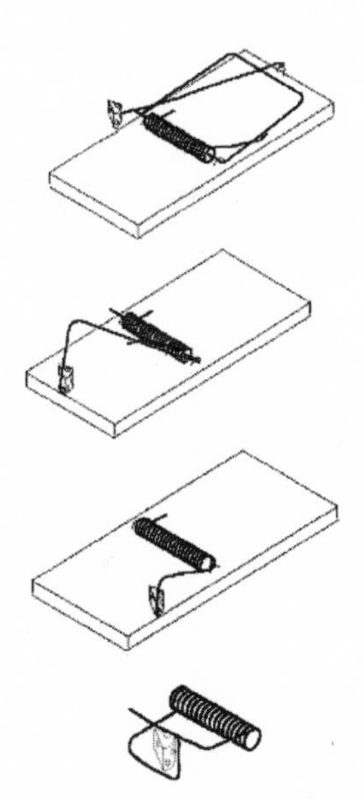

ALTHOUGH the fundamentalist movement went into decline after the Scopes trial, it never vanished. In the 1990s, a new group of evangelical Christians hostile to evolution took over the crusade. The rallying post was *Darwin on Trial* by Berkeley law professor Phillip Johnson, a work in which the author of the *Origin* was tried, convicted, and led away in chains.

As the decade matured, the fundamentalists switched from purely negative attacks on Darwinism to more positive claims. Ostensibly keeping their theological motives to a minimum, this group now directed its attention toward what they called an Intelligent Designer. The biochemist Michael Behe, in his best-selling book *Darwin's Black Box,* most fully articulated the empirical case. He focused on "irreducible complexity," which he defined as "a single system composed of several well-matched, interacting parts that contribute to the basic function, wherein the removal of any one of the parts causes the system to effectively cease functioning" (p. 39). Behe added that any "irreducibly complex biological system, if there is such a thing, would be a powerful challenge to Darwinian evolution. Since natural selection can only choose systems that are already working, then if a biological system cannot be produced gradually it would have to arise as an integrated unit, in one fell swoop, for natural selection to have anything to act on" (p. 39).

As an example of something irreducibly complex, Behe instanced the common mousetrap. This has various parts (five in a standard model—spring, base, and so forth), which are put together to produce a fatal snapping motion when the trigger is activated by a small rodent attempting to take the bait. Behe's point was that this mousetrap is an all-or-nothing phenomenon. Take away any one part and you have something that simply does not function at all. You could not have a mousetrap with four and a half parts or only four parts. "If the hammer were gone, the mouse could dance all night on the platform without becoming pinned to the wooden base. If there were no spring, the hammer and platform would jangle loosely, and again the rodent would be unimpeded" (p. 42). And so on, through the various parts. The mousetrap functions only when it is entirely up and running. It could not have come about through gradual development—and of course, as we know, it did not. It came about through the conscious intent and actions of a human designer. It was planned and fabricated.

Now turn to the world of biology, and in particular turn to the microworld of the cell and of mechanisms that we find at that level. Take bacteria, which use a flagellum, driven by a kind of rotary motor, to move around. Every part is incredibly complex, and so are the various parts when combined. The external filament of the flagellum, for instance, is a single protein which makes a kind of paddle surface that contacts the liquid during swimming. Near the surface of the cell, one finds a thickening—just as needed, so that the filament can be connected to the rotor drive. This naturally requires a connector, known as a "hook protein." There is no motor in the filament, so that has to be somewhere else—at the base of the flagellum, according to Behe, where electron microscopy reveals several ring structures. All way too complex to have come into being gradually, in Behe's view. Only a one-step process would do, and this one-step process must have involved some sort of designing cause. Behe was careful not to identify this designer with the Christian God, but his implication was that it is a force from without the normal course of nature. Things happen through "the guidance of an intelligent agent" (p. 96). More recently Behe has made his own commitment to the Christian God more explicit.

The Conceptual Argument for Intelligent Design

Backing Behe's empirical argument was a conceptual argument presented by the philosopher-mathematician William Dembski. His aim was three-fold. First, to give us the criteria by which we can distinguish something that we would label "designed" rather than otherwise. Second, to show how we distinguish design from something produced naturally by law or something we would put down to chance. Third, to explain why it is impossible for any naturalistic process, including evolution by natural selection, to produce the organic world as we see around us today, that is, a world of design (or, not to prejudge the issue, design appearance).

To infer design, Dembski required three components: contingency, complexity, and specification. Contingency is the idea that something has happened that cannot be ascribed simply to blind law. Being hanged for murder is contingent; falling to the ground when I jump off a stool is not. Design requires contingency. The example Dembski used was the message received from outer space in the movie *Contact*. The series of dots and dashes, zeros and ones, could not be deduced from the laws of physics. They were contingent. But do they show evidence of design?

Here is where complexity comes in, according to Dembski. Suppose we can interpret the series in a binary fashion, and the initial yield is the number group 2, 3, 5. As it happens, these are the beginning of the prime number series, but with so small a yield no one is going to get very excited. It could just be chance, not design. But suppose now you keep going in the series, and it turns out that it yields, in exact and precise order, the prime numbers up to 101. At that point you would start to think that something is up, because the situation seems just too complex to be mere chance. It is highly improbable.

Although most of us would be happy to conclude, on the basis of the improbable prime number sequence, that there are extraterrestrials out there, in fact another component is needed: specification. "If I flip a coin 1000 times, I will participate in a highly complex (that is, highly improbable) event. Indeed, the sequence I end up flipping will be one in a trillion trillion trillion . . . , where the ellipsis needs twenty-two more 'trillions.' This sequence of coin tosses will not, however, trigger a design inference. Though complex, this sequence will not exhibit a suitable pattern." Here, we have a contrast with the prime number sequence from 2 to 101. "Not only is this sequence complex, but it also embodies a suitable pattern. The SETI researcher in the movie *Contact* who discovered this sequence put it this way: 'This isn't noise, this has structure'" (Dembski 2000, 27–28).

What is going on here? According to Dembski, we recognize in design something that is not just arbitrary or random but rather something that was or could be in some way specified, insisted upon, at the outset. We know or could work out the sequence of prime numbers at any time before or after contact from space. The random sequence of penny tosses will come only after the event. "The key concept is that of 'independence.' I define a specification as a match between an event and an independently given pattern. Events that are both highly complex and specified (that is, that match an independently given pattern) indicate design."

This move puts Dembski in a position to go on to the second part of his argument, where we actually detect design. Here we have what he called an explanatory filter. We have a particular phenomenon; the question is, what caused it? Is it something that might not have happened, given the laws of nature? Is it contingent? Or was it necessitated? The moon goes endlessly around the earth. We know that it does this because of Newton's laws. End of discussion. No design here. However, now we have some rather strange new phenomenon, the causal origin of which is a puzzle. Suppose we have a mutation, where although we can predict occurrences over large numbers, we cannot predict a specific individual occurrence. Let us say, as supposedly happened in the extended royal family of Europe, there was a mutation to a gene, and this mutation was responsible for hemophilia. Is it complex? Obviously not, for it leads to breakdown rather than otherwise. Hence it is appropriate to talk here of chance rather than design. The hemophilia mutation was just an accident.

Suppose now that we have a complex mineral pattern where veins of precious metals are set in other materials, the whole being intricate and varied—certainly not a pattern one could simply deduce from the laws of physics or chemistry or geology or whatever. Nor would one think of it as being a breakdown mess, as one might a malmutation. Is this therefore design? Almost certainly not, for there is no way that one might prespecify such a pattern. It is all a bit ad hoc and not something that comes across as the result of conscious intention.

But suppose finally that we have the microscopical biological apparatuses and processes discussed by Behe. They are contingent, they are irreducibly complex, and they are of pre-specified form. As such, they can survive the explanatory filter and be properly considered the product of real design.

Of course, the nature of their designer is another matter. Because one accepts Intelligent Design, one is certainly not thereby necessarily pushed into a literalistic reading of the early chapters of Genesis—six days, six thousand years, universal flood. Still, Dembski was explicit in his belief that evolution is ruled out by this analysis. Why is this? This is where the third, more recent, part of his argument kicked in. Dembski invoked what are known as "no free lunch" theorems, which claim essentially that you cannot get out what you do not put in. Garbage in, garbage out. Or in our specific case: No design in, no design out. In Dembski's opinion, even a guided "theistic" evolution of the Asa Gray variety is unacceptable.

Thinking of an "evolutionary algorithm" as some kind of automatic selection process or rule (like "choose black over white two times out of three"), Dembski writes: "The No Free Lunch theorems underscore the fundamental limits of the Darwinian mechanism. Up until their proof, it was thought that because the Darwinian mechanism could account for all of biological complexity, evolutionary algorithms (i.e., their mathematical underpinnings) must be universal problems solvers. The No Free Lunch theorems show that evolutionary algorithms, apart from careful fine-tuning by a programmer, are no better than blind search and thus no better than pure chance. Consequently, these theorems cast doubt on the power of the Darwinian mechanism to account for all of biological complexity" (Dembski 2002, 212).

Intelligent Design Criticized

Irreducible complexity supposedly could not have come about through unbroken law, and especially not through the agency of natural selection, according to Behe and Dembski. Behe's example of a mousetrap is somewhat unfortunate, however, for it is simply not the case that the trap will work only with all five pieces in place. For a start, one could reduce the number to four by removing the base and fixing the trap to the floor. It might be better if you could move it around, but selection never claimed to produce perfection—simply to produce something that functions better than the immediate alternatives. In fact, as can be seen from the drawing on page 313, it may even be possible to reduce the number of components down to one! Not a great trap, but a trap nevertheless.

Rather more significant, though, is that Behe's mousetrap shows a misunderstanding of the way natural selection works. No Darwinian would deny that in organisms there are parts which, if removed, would lead at once to the malfunctioning or nonfunctioning of the systems in which they occur. The point is not whether the parts now in place could be removed without collapse but whether they could have been put in place by natural selection.

To counter Behe's artifactual analogy, we can think of other artifactual analogies that show precisely how the apparently impossible could be achieved. Consider an arched bridge made from cut stone, without cement, held in place only by the force of the stones against one another. If you tried to build the bridge from scratch, upward and then inward, you would fail—the stones would keep falling to the ground. Indeed, the whole bridge now would collapse if you were to remove the keystone or any stone surrounding it. What you must do to construct an arch is, first, build a supporting structure (possibly an earthen embankment) on which you lay the stones of the bridge, until they are all in place. Then you remove the structure, which is now no longer needed and in fact is in the way of walking under the bridge. Likewise, one can imagine a biochemical sequential process with several stages, on parts of which other processes piggyback, as it were. When the hitherto nonsequential parasitic processes link up and start functioning independently, the original sequence is eventually removed by natural selection because it is redundant or drains resources.

Moving from the pretend to the actual, today's Darwinians have many examples of the most complex of processes that have been put in place by selection. Take as an example the process whereby energy from food is converted into a form that can be used by the cells. This process, called the Krebs cycle, occurs in the cell's mitochondria and centers on two molecules, ATP (adenosine triphosphate) and ADP (adenosine diphosphate). The former is more energy-rich than the latter and is degraded by the body when it needs power. Not only are the molecules themselves complex but so also is the Krebs cycle that makes ATP from other sources of energy. There are almost a dozen sub-processes in the cycle, each making a new product from an earlier one. Every one of these sub-processes demands its own enzyme (these are molecular substances that prime chemical reactions). Do not think that the cycle just appeared out of thin air, complete and entire as it were. Each part of the cycle started life doing something else and then (in true Heath Robinson fashion) was grabbed by the cells and put to a new use. Although, when they set out, the scientists who found all of this out certainly did not have Behe and his "irreducible complexity" in mind, one could imagine that they did, especially from the way in which they set up the problem: "The Krebs cycle has been frequently quoted as a key problem in the evolution of living cells, hard to explain by Darwin's natural selection: How could natural selection explain the building of a complicated structure in toto, when the intermediate stages have no obvious fitness functionality?" (Meléndez-Hervia et al. 1996, 302).

A Behe-type problem, but no Behe-type answer. To the contrary, the various parts of the Krebs cycle had their original uses, and then these were taken over for the whole (p. 302):

> The Krebs cycle was built through the process that Jacob (1977) called "evolution by molecular tinkering," stating that evolution does not produce novelties from scratch: It works on what already exists. The most novel result of our analysis is seeing how, with minimal new material, evolution created the most important pathway of metabolism, achieving the best chemically possible design. In this case, a chemical engineer who was looking for the best design of the process could not have found a better design than the cycle which works in living cells.

Behe's position does not seem plausible, given what we know of the nature of mutation and the stability of biological systems over time. When exactly does the Intelligent Designer supposedly strike to do its work? In *Darwin's Black Box* Behe suggests that everything might have been done long ago and then left to its own devices. "The irreducibly complex biochemical systems that I have discussed . . . did not have to be produced recently. It is entirely possible, based simply on an examination of the systems themselves, that they were designed billions of years ago and that they have been passed down to the present by the normal processes of cellular reproduction" (pp. 227–228). Although Behe ignores the history of the preformed genes from the point between their origin (when they would not have been needed) and today (when they are in full use), we should not. According to the biochemist Kenneth Miller (1999, 162–163), "As any student of biology will tell you, because those genes are not expressed, natural selection would not be able to weed out genetic mistakes. Mutations would accumulate in these genes at breathtaking rates, rendering them hopelessly changed and inoperative hundreds of millions of years before Behe says that they will be needed." There is a mass of experimental evidence showing this to be the case. Behe's idea of a designer doing everything back then and leaving matters to their natural fate is "pure and simple fantasy."

And what about malmutations? If a designer is needed and available to solve complex engineering problems, why could not the designer take some time on the simple matters, specifically those simple matters that if unfixed lead to absolutely horrendous outcomes? Some of the worst genetic diseases are caused by one little alteration in one little part of the DNA. Sickle cell anemia, for example. If the designer is able and willing to do the very complex because it is very good, why does it not do the very simple because the alternative is very bad? Behe speaks of this as being part of the problem of evil—true enough, but just labeling it is not very helpful. Given that the opportunity and ability to do good was so obvious and yet not taken, we need to know why not.

Explanatory Filters

Here Dembski comes to the rescue. A malmutation would surely get caught by the explanatory filter half-way down. It would be siphoned off to the side as chance, if not indeed simply put down as necessity. It certainly would not pass the specification test. This would mean that sickle cell anemia would not be credited to the designer, whereas the flagellum would be. Dembski stressed that these are mutually exclusive alternatives. "To attribute an event to design is to say that it cannot plausibly be referred to either law or chance. In characterizing design as the set-theoretic complement of the disjunction law-or-chance, one therefore guarantees that these three modes of explanation will be mutually exclusive and exhaustive" (Dembski 1998, 98).

To which one might respond that of course one can define things as one will, and if one stipulates that design and law and chance are mutually exclusive, then so be it. But the downside is that one now has a stipulative definition and not necessarily a lexical definition, that is, one which accords to general use. Suppose that something is put down to chance. Does this mean that law is ruled out? Surely not! If I argue that a Mendelian mutation is chance, what I mean is that with respect to that particular theory it is chance. But I may well believe (I surely will believe) that the mutation came about by normal causes and that if these were all known, then it would no longer be chance at all but necessity. As an eminent Victorian scientist once said about these sorts of things: "I have hitherto sometimes spoken as if the variations—so common and multiform in organic beings under domestication, and in a lesser degree in those in a state of nature—had been due to chance. This, of course, is a wholly incorrect expression, but it serves to acknowledge plainly our ignorance of the cause of each particular variation" (Darwin 1859, 131). The point is that chance is a confession of ignorance, not (as one might well think the case in the quantum world) an assertion about the way things are. That is, claims about chance are not ontological assertions, as presumably claims about designers must be.

More than this, one might well argue that the designer always works through law. This takes us back to deism, or to the kind of position endorsed by Baden Powell. The designer may prefer to have things put in motion in such a way that its intentions unfurl and reveal themselves as time goes by. Hugh Miller's pattern in a piece of cloth made by machine is as much an object of design as the pattern of cloth produced by a hand loom. In other words, in a sense that would conform to the normal usage of the terms, one might want to say of something that it is produced by laws, is chance with respect to our knowledge or theory, and fits into an overall context of design by the great orderer or creator of things.

One finds, indeed, real people who have made precisely this kind of claim. At the time he was writing the *Origin,* this probably even applied to Darwin. Later, Ronald Fisher certainly fit the bill. He argued that natural selection made for ongoing progress in evolution. Selection worked on the multiplicity of chance mutations that occur in every natural population. And he wrapped the whole picture up and saw it as the manifestation of the actions and intentions of a good God, in fact (no surprise here) the very Anglican God that Fisher had worshipped all of his life.

No Free Lunch?

Dembski's filter does not let Behe's designer off the hook. If the designer can make—and rightfully takes credit for—the very complex and good, then the designer could prevent—and by its failure is properly criticized for not preventing—the very simple and awful. The problems in theology are as grim as are those in science. But in a way, the real crux of Dembski's case rests with the third part of his argument, about the sheer impossibility of a natural process doing the job. If that part succeeds, then by a kind of neo-Paleyian argument to the best explanation, Dembski can rightly point out that there must be a solution (and why not his?) to the fact that the designer allows (or cannot prevent) the occurrence of some terrible things.

So it is here—on the possibility issue—that we must focus critical attention, and let us start by admitting that impossibility arguments (often known as *reductio ad absurdum* arguments) are well known in mathematics and logic. Take the famous assertion (attributed to Euclid) that there can be no greatest prime number. Assume that there can be, and call the number P. Then consider the number formed by multiplying P by all of the primes less than P and adding one. That is: $2 \times 3 \times 5 \times 7 \times \ldots P_{-1} \times P + 1$. This number is bigger than P and is either itself a prime or factors down into primes bigger than P. (Hint: Nothing prime less than or equal to P can go into the number without leaving a remainder of 1.) Hence, there can be no greatest prime, and you do not need to spend your life checking if a prime greater than (say) $10^{2876} + 37$ exists. It does.

The point here is whether you can offer some similar proof applicable to the real world, showing that selection cannot produce design. Behe offered an empirical argument which, judged on its merits, we have rejected. Dembski offered a more theoretical argument, claiming that no free lunch theorems show that you cannot get design out unless you put design in. Fortunately, it is not necessary to dig into any sophisticated mathematics here for, as Dembski himself acknowledged, the only real relevance of such theorems is that they point to the obvious—that what comes out must have been put in. After stating that "these theorems cast doubt on the power of the Darwinian mechanism to account for all of biological complexity," Dembski (2002, 212) at once concedes: "Granted, the No Free Lunch Theorems are book-keeping results." And then he tries to cover things by adding: "But book-keeping can be very useful. It keeps us honest about the promissory notes our various enterprises—science being one of them—can and cannot make good. In the case of Darwinism we are no longer entitled to think that the Darwinian mechanism can offer biological complexity as a free lunch."

To which of course the Darwinian will reply that this is all very true but beside the point. The Darwinian is not expecting to get out design. The Darwinian is expecting to get out phenomena with the *appearance* of design—the eye, the hand. And the reason why one can get out these design-like phenomena is that natural selection works on the unorganized phenomena that are put in. Reproduction, mutations, competition, and so forth bring on selection, which brings on adaptations, which are design-like phenomena. The question is not whether design demands design. All can grant that. The question is whether design-like demands design, or if selection can do the job instead. It is here that the impossibility proof, if such there be, must strike.

As it happens, such a proof hovers in the background of all discussions like Dembski's. It holds that random mutation simply cannot produce adaptation, or design-like effects. It starts with the rate of mutation, multiples that by the number of genes carried by the average animal or plant, and comes up with a timescale for the creation of organisms that is billions and billions of years beyond the scales of even the greatest projections for the age of the earth. Powers of ten explode outward uncontrollably like the monsters of science fiction movies. "Obviously . . . a million successive, successful mutational steps, each with a probability of one-half, is almost as inconceivable as the instantaneous chance assemblage of a million components into an integrated whole. The chance of success in this case becomes one out of $(2)^{1,000,000}$, or one out of $(10)^{300,000}$" (Morris et al. 1974, 69). And so on. Natural processes producing adaptation is about as likely as a monkey randomly typing a play by Shakespeare, these antievolutionists have claimed.

Famously, Richard Dawkins has scotched this kind of argument. Selection decimates the ever-increasing randomness factors. Choosing a line from Shakespeare—METHINKS IT IS A WEASEL—Dawkins shows that one can set up a computer program that produces letters (or blanks) randomly in all twenty-eight spaces that this sentence occupies. It can produce UMMK JK CDZZ F ZD DSDSKSM, for instance, or S SS FMCV PU I DDRGLKDXRRDO. If one were working randomly, the chances of producing the target sentence in one try is about 1 in 10,000 million million million million million million against. Just impossible in fact. Now rejig the program so that it remembers a successful move, in this case getting a little closer to the target sentence. With this constraint added into a system for producing random letters, the target can be reached in less than fifty moves. So much for all of that multiplication. "There is a big difference, then, between cumulative selection (in which each improvement, however slight, is used as a basis for future building), and single-step selection (in which each new 'try' is a fresh one)" (Dawkins 1986, 49).

The Intelligent Design people do not give this example much weight, pointing out (correctly) that it has many dis-analogies with real-life situations, beginning with the fact that Dawkins specified the end-point, whereas in nature there is supposedly no such fixed goal. Even a progressionist like Dawkins is not going to say that the Cambrian led inevitably to Englishmen. But Dawkins himself conceded this as soon as he gave his example. He was not trying to simulate nature in every respect. He was simply showing that arguments which multiply randomness to a point beyond possibility can be countered simply by factoring in selection. The real question is whether more sophisticated computer models can show that selective forces can generate genuine complexity—the kind that we associate with real adapted organisms—without having to put such complexity into the pool in the first place. And the answer is that they can.

A beautiful example was devised by the biologist Thomas S. Ray. Tierra —an artificial world he created—lives in Ray's computer; the "organisms" who inhabit it are self-replicating programs of an initially fixed length (80 instructions long). "Having determined the address of its [the organism's] beginning and end, it subtracts the two to calculate its size, and allocates a block of memory of this size for a daughter cell. It then calls the copy procedure which copies the entire genome into the daughter-cell memory, one instruction at a time" (Ray 1996, 121–123). Other than being self-replicating, these creatures (as we might neutrally call them) have no functions built in. They exist in a kind of memory pool, and each creature has a chunk of this memory. They cannot write into the program of other creatures, but they can read the information from other creatures and use this information. Time is limited for each creature, and when it has a block of time, it can use it to self-replicate. It then goes to the end of the queue and has to wait its turn for more time. There is a process (the "reaper") by which creatures get eliminated, but creatures can put off extinction by performing tasks (like reproduction) more efficiently than their fellows. Mutations are introduced into the mix by making the information of the creatures, and their replication, subject to random changes.

What happens when you set things in motion (Ray 1996, 124)?

> Once the soup is full, individuals are initially short-lived, generally reproducing only once before dying; thus, individuals turn over very rapidly. More slowly, there appear new genotypes of size 80, and then new size classes. There are changes in the genetic composition of each size class, as new mutants appear, some of which increase significantly in frequency, sometimes replacing the original genotype. The size classes which dominate the community also change through time, as new size classes appear . . . some of which competitively exclude sizes present earlier. Once the community becomes diverse, there is a greater variance in the longevity and fecundity of individuals.
>
> In addition to the raw diversity of genotypes and genome sizes, there is an increase in the ecological diversity. Obligate commensurable parasites evolve, which are not capable of self-replication in isolated culture, but which can replicate when cultured with normal (self-replicating) creatures. These parasites execute some part of the code of their hosts, but cause them no direct harm, except as competitors. Some potential hosts have evolved immunity to the parasites, and some parasites have evolved to circumvent this immunity.

There are even more exciting results, including the evolution of hyper-parasites that not only reproduce on their own but, when subject to parasitism, turn around and use the parasites' own resources to augment their own well-being. One particularly remarkable new feature has an ability to measure internal length when one of the end points (templates) had gone missing. "These creatures located the address of the template marking their beginning, and then the address of a template in the middle of their genome. These two addresses were then subtracted to calculate half of their size, and this value was multiplied by two (by shifting left) to calculate their full size" (p. 127).

All of this leads to the conclusion that here we have the evolution of adaptations—of specified complexity—through selection, without the need to introduce design in the first place. Just what people like Dembski say cannot happen. Of course this is an artificial situation. We have not got real organisms, simplifications have been introduced, the program was set up to give results in our lifetimes, and so forth. But to object that this (or any similar program) therefore does not prove the nondesigned production of specified complexity, because Ray (or whoever) had to build up the system in the first place, and this required design, is to miss the whole point of experimentation. The key question is whether, even in principle, blind variation acted upon by selection can produce adaptation—design-like features. And the answer is that it can and did.

You cannot get much blinder than Ray's variation. He had no control over where and when it struck. It produced parasitism. Of course it was artificial parasitism. We know that. But if you asked a programmer to produce parasitism, he or she could not get it by striking the keys randomly; design would be required. We know also that an artificial situation is not a real situation. Whether Ray's system tells us anything real about nature requires questions about mutation rate and so forth, and this is another matter. What we are dealing with now is Dembski's claim that blind processes cannot lead to specified complexity, and a system like Ray's Tierra shows that Dembski is just plain wrong.

The Blind Watchmaker

Suppose you decide to take up Darwinian evolution. Does this now commit you to reject natural theology—or a theology of nature? Is Darwinism positively hostile to any kind of genuine design—inexplicable, illustrative, inspirational, or whatever? Dawkins is one of several who have argued just this. Science and religion make claims, and those claims clash. One side is right and one side is wrong. Dawkins gives and expects no quarter. "A cowardly flabbiness of the intellect afflicts otherwise rational people confronted with long-established religions" (Dawkins 1997). As a Darwinian, Dawkins was not about to deny that the Christians got the question right. It was their answer that they got wrong. Speaking of Archdeacon William Paley with respect, if not reverence, Dawkins willingly allowed that *Natural Theology* articulates the teleological argument "more clearly and convincingly than anyone had before" (Dawkins 1986, 4). The trouble is that "the analogy between telescope and eye, between watch and living organism, is false. All appearances to the contrary, the only watchmaker in nature is the blind forces of physics, albeit deployed in a very special way." A genuine watchmaker plans everything and puts his purposes into action. Natural selection simply acts. It has no purposes. "It has no mind and no mind's eye. It does not plan for the future. It has no vision, no foresight, no sight at all. If it can be said to play the role of watchmaker in nature, it is the *blind* watchmaker" (p. 5). To put it in our language, Paley was spot-on about the argument to complexity, but he was dead wrong about the argument to design.

Now obviously not all Christians are going to be upset by this. Wolfhart Pannenberg's response will simply be that Paley's mistake was in thinking that God's creative powers could act only through miracle. (Frederick Temple made exactly this point more than a hundred years ago.) Darwin showed that the route to creation must lie through evolution by natural selection. But why should not God have worked this way rather than in some other fashion? Darwin himself thought this when he wrote the *Origin of Species.* It is true that, by this time, Darwin was a deist rather than a theist, but among Christians the Augustinian for one has no theological worries here. You cannot separate Becoming and Being. Although Augustine himself was certainly no evolutionist, he laid the theological groundwork for precisely such a scientific theory.

None of this will convince Dawkins, who has a number of back-up arguments ready to keep Darwinism and Christianity apart. Most powerful to him as a counter-argument against God is the traditional problem of pain and evil, which the Darwinian approach exacerbated. Natural selection presupposes a struggle for existence, and the struggle on many, many occasions is downright nasty. Using the notion of a "utility function" for the end purpose being intended, Dawkins drew attention to the interactions between cheetahs and antelopes, and asks: "What was God's utility function?" Cheetahs seem wonderfully designed to kill antelopes. "The teeth, claws, eyes, nose, leg muscles, backbone and brain of a cheetah are all precisely what we should expect if God's purpose in designing cheetahs was to maximize deaths among antelopes." But conversely, "we find equally impressive evidence of design for precisely the opposite end: the survival of antelopes and starvation among cheetahs." It is almost as though two warring gods were at work, making different animals and then setting them to compete with one another. If there is only one god who made the two animals, then what kind of god could it be? "Is He a sadist who enjoys spectator blood sports? Is He trying to avoid overpopulation in the mammals of Africa? Is He maneuvering to maximize David Attenborough's television ratings?" The whole thing is ludicrous (Dawkins 1995, 105). Truly, concludes Dawkins, there are no ultimate purposes to life, no deep religious meanings. There is nothing.

Pain

At this point spokespersons in the science/religion community simply retreat. On their way out the door, most agree that Dawkins has a good point—Darwinism does show that life is mean and nasty, without any purpose. But the best way to avoid so unpleasant a conclusion is to abandon Darwinism, rather than give up on religion or even evolution per se. They scramble to find an evolutionism with a warmer, friendlier face. After all, there must be some way to get what you want—adaptation possibly, humans definitely—with a fairly firm guarantee of desired direction. We have seen Holmes Rolston III in action, and he stands out only because—to his credit—he is so clear and forthright in his likes and dislikes and in his personal conviction that Darwinism is not going to speak to them.

This will not do. Whether or not Darwinism—adaptation brought about by natural selection—taken as a whole is both true and in basic outlines essentially complete, it is an active and forward-looking science. Whether we like it or not, we are stuck with it. The Darwinian revolution is over, and Darwin won. Hence, any satisfactory response to Dawkins must be on his terms—adaptation, selection, blind variation, pain, and all. It must be on terms that recognize that Dawkins is right, that Darwinism is a major challenge to religious belief, and that you cannot simply pretend that nothing very much has happened. But once this is accepted, things start to fall into place.

Dawkins is absolutely right that total separation of religion and science is no response. Darwinism *does* talk about origins, and the theologically inclined must take note of what it teaches. Even if one goes the route of Newman by separating science and religion, one must agree that this path does not confer a license to say whatever one likes about events and phenomena in the other domain. Pure Paley is also no longer possible in light of Darwinian evolution; natural selection rules out the necessity of an appeal to an intervening God. This leaves only the third option, a theology of nature. The only possible response to Dawkins is that, Darwin or not, you feel compelled to accept that our understanding of nature, of living things, is changed and illuminated and made complete by your acceptance of the existence and creative power and sustaining nature of God.

Where stands this response in the light of Dawkins's attack? Can one still "relate all of nature to the reality that is the true theme of theology—the reality of God" (Pannenberg 1993, 73)? What about the problem of evil? This is always the really tough one. Dawkins wants to claim that the coming of Darwin made the God hypothesis impossible. After Darwin we see that the world is simply not how it would be if an all-loving, all-powerful creator had made it (the cheetah and antelope argument). Nothing like this could possibly occur, given a loving god. But this is exactly what one would expect from blind, purposeless law. However, the appeal to blind unbroken law—things not working perfectly and pain and strife being commonplace—can be turned on its head, to yield a traditional argument protecting the possibility of the Christian God. As Darwin himself saw, one must and can properly invoke some version of the argument used by the philosopher Leibniz, namely that God's powers only extend to doing the possible. Having once committed himself to creation by law, everything else follows as a matter of course. Leibniz's argument was parodied by Voltaire in *Candide*—everything in this world happens for the best of all possible reasons—but just as that counter-argument fails in the world of religion, so also it fails in the world of science.

Think of trying to make physical precesses entirely pain-free. Start with fire. It could no longer burn or produce smoke. But if this were the case, would it still be hot? If so, then how could this be? There would have to be wholesale changes at the molecular level. If fire were not hot, how would we warm ourselves and cook our food? One change by God would require a knock-on compensation, and another and another. And even then, could we get a system that worked properly? "Human beings are sentient creatures of nature. As physiological beings they interact with Nature; they cause natural events and in turn are affected by natural events. Hence, insofar as humans are natural, sentient beings, constructed of the same substance as Nature and interacting with it, they will be affected in any natural system by lawful natural events" (Reichenbach 1982, 111–112).

Once we start altering things to eliminate pain and suffering, there is no end to it—except that we would have changed humans so that they are no longer truly human. And even then, who would dare say we humans would be better situated? "Whether humans would have evolved but no infectious virus or bacilli, or whether there would have resulted humans with worse and more painful diseases, or whether there would have been no conscious, moral beings at all, cannot be discerned" (p. 113). All things in the world fit into an interlocking whole. We should not assume that God could have made things in another way, avoiding instances of physical evil.

In fact, somewhat humorously, one can appeal to Dawkins's own writings to drive home this line of argument. As a good Darwinian, he insists that adaptive complexity is the mark of the living. This and this alone picks out the quick from the dead, or rather from the never quick. But how is one to get this adaptive complexity? Dawkins insists that it is through and only through natural selection. Alternative mechanisms like Lamarckism or hopeful monsters (mutationism) simply will not work. "Wherever in the universe adaptive complexity shall be found, it will have come into being gradually through a series of small alterations, never through large and sudden increments in adaptive complexity" (Dawkins 1983, 412). Adaptive complexity does not come through regular physical processes, immediately bringing on marks of the living. Sudden changes bring chaos and disorder. Planes crashing break up into useless pieces. Planes do not suddenly and spontaneously rise from the junkyard, pieces coming together into functioning wholes. Likewise for organisms. Dawkins has no love of Stuart Kauffman's "order for free." There is no alternative to natural selection. No other purely physical process brings about the adaptive, organized complexity of living things. "However diverse evolutionary mechanisms may be, if there is no other generalization that can be made about life all around the Universe, I am betting that it will always be recognizable as Darwinian life. The Darwinian Law . . . may be as universal as the great laws of physics" (Dawkins 1983, 423). You cannot get adaptive complexity without natural selection.

In other words, if God's process of creation is through unbroken law, then he had to do it as he did—natural selection, pain and agony, imperfection, and all. One might still argue that God should not have created in the first place (it is an essential part of Christian theology that the act of creation was done freely from love by God), but that is a somewhat different point. If one objects, as did the novelist Dostoevsky, that the pain in the world could never be trumped by the happy end result—no amount of bliss in heaven for Mother Teresa could ever balance the agony of a neglected, starving child—one is not necessarily implicating the coming of Darwinism. The pain would happen irrespective of natural selection. So, ultimately, effective though this line of argument may or may not be, it is irrelevant to our main line of discussion. Darwinism as such does not make irresistible the argument from natural evil. It may even make the solution easier to grasp.

Turning Back the Clock?

Behe, Dembski, and their nemesis, Dawkins, share a desire to return to the high Victorian era, when Britain ruled the waves and science and religion could never agree. They would have made sure that the clash between Huxley and Wilberforce really was everything claimed of it. But the world has moved on. The old disjunction is gone. Evolution has been proven true, and is widely accepted as such. God did not intervene miraculously to make each species separately. Christians can and do accept this fact. Yet, in another sense, I come to praise these old-fashioned people and to argue that they have grasped something from the past that too many who write on the science/religion relationship have lost or never even seen. This is the argument to organized, adaptive complexity.

However one might criticize Behe's conclusions, when he speaks about the inner workings of the cell, his audience senses the presence of a man who truly loves the natural world. Say what you like in criticism of Dawkins, when he writes about the echolocation mechanism of the bat or about the eye and its varieties, he reveals to his readers an uncommon delight in the intricate workings of the organic world. In this Behe and Dawkins are at one with Aristotle, John Ray, Georges Cuvier, and of course Charles Darwin. And they are at one with modern-day researchers like Nicholas Davies, who spent long hours, tedious in the extreme to an outsider, standing in the dark and the rain of an early Cambridge morning with his dunnocks and cuckoos—observing, measuring, theorizing. Scrabbling around in the muck of the undergrowth to find just what it was that the female dunnock ejected when the male pecked at her backside. Patiently holding back as the cuckoo chick systematically destroys all of its foster siblings. Hauling off to Iceland with phony, plastic eggs to test a hypothesis. A man more in touch—literally as well as metaphorically—with the living world, and in love with it, we would have difficulty imagining. But he is equaled by all of the others, crouched over their microscopes, marveling at the beauty of the eyes of the fruit fly; sweating in the humid tropical heat, waist-deep in water searching for little fish; clambering around fig trees in Panama trying to find minute wasps. All of these people are simply overwhelmed by the glory of creation. They appreciate the organized complexity of the natural world.

Around 1875, natural theology took a wrong if understandable turn, and in the process lost this appreciation. Natural theology threw out the baby with the bathwater, for with the old-fashioned argument to design it rejected—or at least started significantly to ignore—the argument to complexity. It thought that when Paley's conclusion went, his premises had to go too. This, again, was understandable around 1875, when Darwinians like Huxley were stressing homology and belittling adaptation. But this rejection of the argument to complexity was simply a mistake. Adaptation reigns as never before, and it moves people as much as it ever did—more in fact, because now we can start to understand the why as well as the what. Gilbert White knew about the dunnocks and cuckoos, but how much deeper and more profound and more satisfying is the understanding of Nicholas Davies. Cope and Marsh knew about stegosaurus, but how much deeper and more profound and more satisfying is the understanding of those who have worked out the reasons for those ridiculous plates. Probably nobody knew much about fruit flies, and yet see how they have opened up the mysteries of nature to evolutionists.

What I am arguing for is a theology of nature—for let us agree that natural theology is now gone—where the focus is back on adaptation. A theology of nature that sees and appreciates the complex, adaptive glory of the living world, rejoices in it, and trembles before it. I argue for this even though the people who reveal it to us today in its fullest majesty may be people for whom Christianity evokes emotions ranging from bored indifference to outright hostility. This is irrelevant, especially to those of us who know professional Darwinian evolutionists. As Ernst Mayr once said to me: "People forget that it is possible to be intensely religious in the entire absence of theological belief" (interview, March 30, 1988). Theologians working on the science/religion relationship, few of whom have actually had hands-on experience with nature, let the hostility of atheists like Dawkins, or their embarrassment with the Intelligent Design enthusiasts, blind them to the genuine love and joy with which today's professional evolutionists respond to their subjects. We should strip away the pseudo-arguments in the way of a full appreciation of the argument to complexity and start sharing what moved the natural theologians of old and what still moves the evolutionists of today.

Canon Charles Raven, although no friend to Darwinism, was a naturalist as well as a theologian. He knew whereof I speak. In his Gifford Lectures, he wrote of how much time he had spent and sheer pleasure he had derived from following and studying butterflies all over England and Scotland. "Every specimen differed from the rest, in detail from those of its own group, in total effect from those of others. Each was in itself a perfect design, satisfying in whole and parts, inviting one to concentrate one's whole attention upon it. To move from one to another, to sense the difference of impact, to work out the quality of this difference in the detailed modifications of the general pattern, this was a profoundly moving experience." For Raven, this was the real edge of the science/religion encounter. This is what makes it all meaningful to the believer. Not proof, but simply flooding, overwhelming experience that could not be denied. In Raven's words (1953, 112–113): "Here is beauty—whatever the philosophers and art critics who have never looked at a moth may say—beauty that rejoices and humbles, beauty remote from all that is meant by words like random or purposeless, utilitarian or materialistic, beauty in its impact and effects akin to the authentic encounter with God."

I have nothing more to add.

SOURCES AND SUGGESTED READING

Introduction

Aristotle. 1984. Physics. *The Complete Works of Aristotle.* Ed. J. Barnes. Princeton: Princeton University Press.
Darwin, C. 1859. *On the Origin of Species.* London: John Murray.
Dawkins, R. 1986. *The Blind Watchmaker.* New York: Norton.
Kershaw, I. 1999. *Hitler, 1889–1936: Hubris.* New York: Norton.
Mackie, J. 1966. The direction of causation. *Philosophical Review* 75: 441–466.
Williams, G. C. 1966. *Adaptation and Natural Selection.* Princeton: Princeton University Press.

1. Two Thousand Years of Design

Aquinas, St. T. 1952. *Summa Theologica,* 1. London: Burns, Oates and Washbourne.
Augustine. 1998. *The City of God against the Pagans.* Ed. and trans. R. W. Dyson. Cambridge: Cambridge University Press.
Bacon, F. [1605] 1868. *The Advancement of Learning.* Oxford: Clarendon Press.
Barnes, J., ed. 1984. *The Complete Works of Aristotle,* 1. Princeton: Princeton University Press.
Boyle, R. 1996. *A Free Enquiry into the Vulgarly Received Notion of Nature.* Eds. E. B. Davis and M. Hunter. Cambridge: Cambridge University Press.
Calvin, J. 1962. *Institutes of the Christian Religion.* Grand Rapids: Eerdmans.
Cooper, J. M., ed. 1997. *Plato: Complete Works.* Indianapolis: Hackett.
Dawkins, R. 1983. Universal Darwinism. In *Molecules to Men,* ed. D. S. Bendall. Cambridge: University of Cambridge Press.
Descartes, R. 1985. *The Philosophical Writings,* 1. Trans. J. Cottingham, R. Stoothoff, and D. Murdoch. Cambridge: Cambridge University Press.

Galen. 1968. *On the Usefulness of the Parts of the Body.* Trans. M. T. May. Ithaca: Cornell University Press.

Gotthelf, A., and J. G. Lennox, eds. 1987. *Philosophical Issues in Aristotle's Biology.* Cambridge: Cambridge University Press.

Goudge, T. 1973. Evolutionism. In *Dictionary of the History of Ideas.* New York: Scribner's.

Hume, D. [1779] 1947. *Dialogues Concerning Natural Religion.* Ed. N. K. Smith. Indianapolis: Bobbs-Merrill.

Hurlbutt, R. H. 1965. *Hume, Newton, and the Design Argument.* Lincoln: University of Nebraska Press.

Lucretius. 1969. *The Way Things Are: The De Rerum Natura of Titus Lucretius Carus.* Trans. R. Humphries. Bloomington: Indiana University Press.

McGrath, A. 2001. *A Scientific Theology,* 1: *Nature.* Grand Rapids: Eerdmans.

McPherson, T. 1972. *The Argument from Design.* London: Macmillan.

Müller, E. F. K., ed. 1903. *Die Bekenntnissschriften der Reformierten Kirche.* Leipzig: Böhme.

2. Paley and Kant Fight Back

Boyle, R. [1688] 1966. A disquisition about the final causes of natural things. In *The Works of Robert Boyle,* 5, ed. T. Birch. Hildesheim: Georg Olms.

Brigden, S. 2001. *New Worlds, Lost Worlds: The Rule of the Tudors, 1485–1603.* New York: Viking.

Buckland, W. 1836. *Geology and Mineralogy.* Bridgewater Treatise, 6. London: William Pickering.

Butler, J. [1736] 1824. *The Analogy of Religion, Natural and Revealed, to the Constitution and Course of Nature.* London: Longman, Hurst, Rees, Orme.

Cannon, W. F. 1978. *Science in Culture: The Early Victorian Period.* New York: Science History Publications.

Derham, W. 1716. *Physico-Theology.* London: Innys.

Fermat, P. de. 1891–1912. *Oeuvres de Fermat.* Eds. P. Tannery and C. Henry. Paris: Gauthier-Villars.

Gillespie, N. C. 1987. Natural history, natural theology, and social order: John Ray and the "Newtonian Ideology." *Journal of the History of Biology* 20: 1–49.

Gordon, E. O. 1894. *The Life and Correspondence of William Buckland.* London: John Murray.

Hooker, R. 1845. *The Collected Works.* Oxford: Oxford University Press.

Hume, D. [1779] 1947. *Dialogues Concerning Natural Religion.* Ed. N. K. Smith. Indianapolis: Bobbs-Merrill.

Kant, I. [1790] 1928. *The Critique of Teleological Judgement.* Trans. J. C. Meredith. Oxford: Oxford University Press.

———. [1781] 1929. *Critique of Pure Reason*. Trans. N. Kemp Smith. New York: Humanities Press.
McFarland, J. D. 1970. *Kant's Concept of Teleology*. Edinburgh: University of Edinburgh Press.
Newton, I. 1782. *Opera Quae Exstant Omnia*. Ed. Horsely. London.
Olson, R. 1987. On the nature of God's existence, wisdom, and power: the interplay between organic and mechanistic imagery in Anglican natural theology—1640–1740. In *Approaches to Organic Form: Permutations in Science and Culture*, ed. F. Burwick. Dordrecht: Reidel.
Osler, M. J. 2001. Whose ends? Teleology in early modern natural philosophy. *Osiris* 16: 151–168.
Pagel, W. 1967. *William Harvey's Biological Ideas: Selected Aspects and Historical Background*. Basel: Karger.
Paley, W. [1794] 1819. *Collected Works, 3: Evidences of Christianity*. London: Rivington.
———. [1802] 1819. *Collected Works, 4: Natural Theology*. London: Rivington.
Ray, J. 1709. *The Wisdom of God, Manifested in the Works of Creation*, 5th ed. London: Samuel Smith.
Sober, E. 2000. *Philosophy of Biology*, 2nd ed. Boulder: Westview.
Zammito, J. H. 1992. *The Genesis of Kant's Critique of Judgment*. Chicago: University of Chicago Press.

3. *Sowing the Seeds of Evolution*

Appel, T. A. 1987. *The Cuvier-Geoffroy Debate: French Biology in the Decades before Darwin*. New York: Oxford University Press.
Chambers, R. 1844. *Vestiges of the Natural History of Creation*. London: Churchill.
Cuvier, G. 1813. *Essay on the Theory of the Earth*. Trans. Robert Kerr. Edinburgh: W. Blackwood.
———. 1817. *Le règne animal distribué d'après son organisation, pour servir de base à l'histoire naturelle des animaux et d'introduction à l'anatomie comparée*. Paris: Déterville.
———. 1858. *Lettres de Georges Cuvier à C.H. Pfaff, 1788–1792*. Paris: Masson.
Darwin, E. 1801. *Zoonomia; or, The Laws of Organic Life*, 3rd ed. London: J. Johnson.
Geoffroy Saint-Hilaire, E. 1818. *Philosophie anatomique*. Paris: Mequignon-Marvis.
Goethe, J. W. 1946. On the metamorphosis of plants. In *Goethe's Botany*, ed. A. Arber. Waltham: Chronica Botanica.
Kant, I. [1790] 1928. *The Critique of Teleological Judgement*. Trans. J. C. Meredith. Oxford: Oxford University Press.
King-Hele, D. 1981. *The Letters of Erasmus Darwin*. Cambridge: Cambridge University Press.

Lamarck, J. B. 1809. *Philosophie zoologique.* Paris: Dentu.
Letteney, M. J. Georges Cuvier, transcendental naturalist: a study of teleological explanation in biology. Ph.D. diss., University of Notre Dame.
Lovejoy, A. O. [1911] 1959. Kant and evolution. In *Forerunners of Darwin,* eds. B. Glass, O. Temkin, and W. L. Strauss Jr. Baltimore: Johns Hopkins University Press.
McMullin, E. 1985. Introduction: evolution and creation. In *Evolution and Creation,* ed. E. McMullin. Notre Dame: University of Notre Dame Press.
Powell, B. 1855. *Essays on the Spirit of the Inductive Philosophy.* London: Longman, Brown, Green, and Longman.
Richards, R. J. 2000. Kant and Blumenbach on the "Bildungstrieb": A historical misunderstanding. *Studies in History and Philosophy of Biological and Biomedical Sciences* 31: 11–32.
Ruse, M. 1996. *Monad to Man: The Concept of Progress in Evolutionary Biology.* Cambridge: Harvard University Press.
Wells, G. A. 1967. Goethe and evolution. *Journal of the History of Ideas* 28: 537–550.

4. *A Plurality of Problems*

Babbage, C. [1838] 1967. *The Ninth Bridgewater Treatise. A Fragment.* 2nd ed. London: Frank Cass.
Bebbington, D. W. 1999. Science and evangelical theology in Britain from Wesley to Orr. In *Evangelicals and Science in Historical Perspective,* eds. D. N. Livingstone, D. G. Hart, and M. A. Noll. Oxford: Oxford University Press.
Brewster, D. 1854. *More Worlds than One: The Creed of the Philosopher and the Hope of the Christian.* London: Camden Hotten.
Chalmers, T. [1817] 1906. *A Series of Discourses on the Christian Revelation, Viewed in Connection with the Modern Astronomy.* New York: American Tract Society.
Clark, J. W., and T. M. Hughes, eds. 1890. *Life and Letters of the Reverend Adam Sedgwick.* Cambridge: Cambridge University Press.
Ker, I. 1989. *John Henry Newman: A Biography.* Oxford: Oxford University Press.
Kimler, W. 1983. One hundred years of mimicry: history of an evolutionary exemplar. Ph.D. diss., Cornell University.
Kirby, W., and W. Spence. 1815–1828. *An Introduction to Entomology: or Elements of the Natural History of Insects.* London: Longman, Hurst, Reece, Orme, and Brown.
Miller, H. 1856. *The Testimony of the Rocks, or Geology in Its Bearings on the Two Theologies, Natural and Revealed.* Edinburgh: Constable.
Newman, J. H. 1870. *A Grammar of Assent.* New York: Catholic Publishing Society.

———. 1973. *The Letters and Diaries of John Henry Newman,* 25. Eds. C. S. Dessain and T. Gornall. Oxford: Clarendon Press.

Owen, R. 1834. On the generation of the marsupial animals, with a description of the impregnated uterus of the kangaroo. *Philosophical Transactions*: 333–364.

———. 1848. *On the Archetype and Homologies of the Vertebrate Skeleton.* London: Voorst.

Owen, Rev. R. 1894. *The Life of Richard Owen.* London: Murray.

Rupke, N. A. 1994. *Richard Owen: Victorian Naturalist.* New Haven: Yale University Press.

Todhunter, I. 1876. *William Whewell, DD. Master of Trinity College Cambridge: An Account of His Writings with Selections from His Literary and Scientific Correspondence.* London: Macmillan.

Topham, J. R. 1999. Science, natural theology, and evangelicalism in early nineteenth-century Scotland: Thomas Chalmers and the *Evidence* controversy. In *Evangelicals and Science in Historical Perspective,* eds. D. N. Livingstone, D. G. Hart, and M. A. Noll. Oxford: Oxford University Press.

Whewell, W. 1833. *Astronomy and General Physics.* Bridgewater Treatise, 3. London.

———. 1837. *The History of the Inductive Sciences.* London: Parker.

———. 1840. *The Philosophy of the Inductive Sciences.* London: Parker.

———. 1845. *Indications of the Creator.* London: Parker.

———. [1853] 2001. *Of the Plurality of Worlds. A Facsimile of the First Edition of 1853: Plus Previously Unpublished Material Excised by the Author Just before the Book Went to Press; and Whewell's Dialogue Rebutting His Critics, Reprinted from the Second Edition.* Ed. M. Ruse. Chicago: University of Chicago Press.

5. *Charles Darwin*

Barrett, P. H., P. J. Gautrey, S. Herbert, D. Kohn, and S. Smith, eds. 1987. *Charles Darwin's Notebooks, 1836–1844.* Ithaca: Cornell University Press.

Browne, J. 1995. *Charles Darwin: Voyaging.* New York: Knopf.

———. 2002. *Charles Darwin: The Power of Place.* New York: Knopf.

Darwin, C. 1842. *The Structure and Distribution of Coral Reefs.* London: Smith Elder.

———. 1859. *On the Origin of Species.* London: John Murray.

———. 1871. *The Descent of Man.* London: John Murray.

Darwin, F., ed. 1887. *The Life and Letters of Charles Darwin, including an Autobiographical Chapter.* London: John Murray.

Darwin, F., and A. C. Seward, eds. 1903. *More Letters of Charles Darwin.* London: John Murray.

Ghiselin, M. T. 1984. Introduction to C. Darwin, *The Various Contrivances by Which Orchids Are Fertilized by Insects.* Chicago: University of Chicago Press.

Huxley, T. H. 1863. *Evidence as to Man's Place in Nature.* London: Williams and Norgate.

———. [1864] 1893. Criticisms on "The Origin of Species." *Darwiniana.* London: Macmillan.

Kottler, M. J. 1974. Alfred Russel Wallace, the origin of man, and spiritualism. *Isis* 65: 145–192.

Lennox, J. G. 1993. Darwin was a teleologist. *Biology and Philosophy* 8: 409–421.

Lyell, C. 1830–1833. *Principles of Geology: Being an Attempt to Explain the Former Changes in the Earth's Surface by Reference to Causes Now in Operation.* London: John Murray.

Malthus, T. R. 1826. *An Essay on the Principle of Population,* 6th ed. London: John Murray.

Ruse, M. 1979. *The Darwinian Revolution: Science Red in Tooth and Claw.* Chicago: University of Chicago Press.

———. 1980. Charles Darwin and group selection. *Annals of Science* 37: 615–630.

6. A Subject Too Profound

Barrett, P. H., P. J. Gautrey, S. Herbert, D. Kohn, and S. Smith, eds. 1987. *Charles Darwin's Notebooks, 1836–1844.* Ithaca: Cornell University Press.

Darwin, C. 1854a. *A Monograph of the Fossil Balanidae and Verrucidae of Great Britain.* London: Palaeontographical Society.

———. 1854b. *A Monograph of the Sub-Class Cirripedia, with Figures of All the Species. The Balanidge (or Sessile Cirripedes); the Verrucidae, and C.* London: Ray Society.

———. 1859. *On the Origin of Species.* London: John Murray.

———. 1862. *On the Various Contrivances by Which British and Foreign Orchids Are Fertilized by Insects, and on the Good Effects of Intercrossing.* London: John Murray.

———. 1868. *The Variation of Animals and Plants under Domestication.* London: John Murray.

———. 1871. *The Descent of Man.* London: John Murray.

———. 1872. *The Expression of the Emotions in Man and Animals.* London: John Murray.

———. 1958. *The Autobiography of Charles Darwin, 1809–1882.* Ed. N. Barlow. London: Collins.

———. 1959. *The Origin of Species by Charles Darwin: A Variorum Text.* Ed. M. Peckham. Philadelphia: University of Pennsylvania Press.

———. 1985–. *The Correspondence of Charles Darwin.* Cambridge: Cambridge University Press.

Darwin, C., and A. R. Wallace. 1958. *Evolution by Natural Selection*. Foreword by G. de Beer. Cambridge: Cambridge University Press.

Darwin, F., and A. C. Seward, eds. 1903. *More Letters of Charles Darwin*. London: John Murray.

Di Gregorio, M., and N. W. Gill, eds. 1990. *Charles Darwin's Marginalia*, 1. New York: Garland.

Henslow, J. S. 1836. *Descriptive and Physiological Botany*. London: Longman, Rees, Orme, Brown, Green, and Longman.

Owen, R. 1992. *The Hunterian Lectures in Comparative Anatomy, May and June 1837*, ed. P. R. Sloan. Chicago: Chicago University Press.

7. *Darwinian against Darwinian*

Bannister, R. 1979. *Social Darwinism: Science and Myth in Anglo-American Social Thought*. Philadelphia: Temple University Press.

Bateson, B. 1928. *William Bateson, F.R.S., Naturalist: His Essays and Addresses Together with a Short Account of His Life*. Cambridge: Cambridge University Press.

Darwin, C. 1868. *The Variation of Animals and Plants under Domestication*. London: John Murray.

———. 1985–. *The Correspondence of Charles Darwin*. Cambridge: Cambridge University Press.

Desmond, A. 1994. *Huxley, the Devil's Disciple*. London: Michael Joseph.

———. 1997. *Huxley, Evolution's High Priest*. London: Michael Joseph.

Dupree, A. H. 1959. *Asa Gray, 1810–1888*. Cambridge: Harvard University Press.

Gray, A. 1876. *Darwiniana*. New York: D. Appleton.

———. 1879. *Structural Botany*. New York: Ivison, Blakeman, Taylor.

Huxley, L., ed. 1900. *The Life and Letters of Thomas Henry Huxley*. London: Macmillan.

Huxley, T. H. 1854. Vestiges, etc. *British and Foreign Medico-Chirurgical Review* 13: 425–439.

———. 1873. *Critiques and Addresses*. New York: Appleton.

———. 1879. *Hume*. London: Macmillan.

———. [1859] 1893. The Darwinian hypothesis. *Times*, December 26. *Darwiniana*. London: Macmillan.

———. [1864] 1893. Criticisms on "The Origin of Species." *Darwiniana*. London: Macmillan.

———. 1903. *The Scientific Memoirs of Thomas Henry Huxley*. Eds. M. Foster and E. R. Lankester. London: Macmillan.

Nyhart, L. K. 1995. *Biology Takes Form: Animal Morphology and the German Universities*. Chicago: University of Chicago Press.

Wright, C. 1871. *Darwinism: Being an Examination of Mr. St. George Mivart's "Genesis of Species."* London: John Murray.

8. The Century of Evolutionism

Bates, H. W. (1862). Contributions to an insect fauna of the Amazon Valley Lepidoptera: *Heliconidae, Transactions of the Linnaean Society of London,* 23, 495–515.

———. [1863] 1892. *The Naturalist on the River Amazon.* London: John Murray.

Bateson, W. 1894. *Materials for the Study of Variation, Treated with Especial Regard to Discontinuity in the Origin of Species.* London: Macmillan.

Bennett, J. H., ed. 1983. *Natural Selection, Heredity, and Eugenics. Including Selected Correspondence of R. A. Fisher with Leonard Darwin and Others.* Oxford: Oxford University Press.

Box, J. F. 1978. *R. A. Fisher: The Life of a Scientist.* New York: Wiley.

Cain, A. J., and P. M. Sheppard. 1954. Natural selection in Cepaea. *Genetics* 39: 89–116.

Castle, W. E., and J. C. Phillips. 1914. *Piebald Rats and Selection: An Experimental Test of the Effectiveness of Selection and of the Theory of Gamete Purity in Mendelian Crosses.* Washington, DC: Carnegie Institution of Washington.

Dobzhansky, T. 1937. *Genetics and the Origin of Species.* New York: Columbia University Press.

Fisher, R. A. 1930. *The Genetical Theory of Natural Selection.* Oxford: Oxford University Press.

———. 1947. The renaissance of Darwinism. *Listener* 37: 1001.

Ford, E. B. 1964. *Ecological Genetics.* London: Methuen.

Goodrich, E. S. 1912. *The Evolution of Living Organisms.* London: T. C. and E. C. Jack.

Lewontin, R. C., J. A. Moore, W. B. Provine, and B. Wallace, eds. 1981. *Dobzhansky's Genetics of Natural Populations.* New York: Columbia University Press.

Mayr, E. 1942. *Systematics and the Origin of Species.* New York: Columbia University Press.

Provine, W. B. 1971. *The Origins of Theoretical Population Genetics.* Chicago: University of Chicago Press.

———. 1986. *Sewall Wright and Evolutionary Biology.* Chicago: University of Chicago Press.

Punnett, R. C. 1915. *Mimicry in Butterflies.* Cambridge: Cambridge University Press.

Sheppard, P. M. 1958. *Natural Selection and Heredity.* London: Hutchinson.

Simpson, G. G. 1944. *Tempo and Mode in Evolution.* New York: Columbia University Press.

Stebbins, G. L. 1950. *Variation and Evolution in Plants.* New York: Columbia University Press.

Vorzimmer, P. J. 1970. *Charles Darwin: The Years of Controversy.* Philadelphia: Temple University Press.

Wallace, A. R. 1866. On the phenomena of variation and geographical distribution as illustrated by the Papilionidae of the Malayan region. *Transactions of the Linnaean Society London* 25: 1–27.

Wright, S. [1931] 1986. Evolution in Mendelian populations. In *Evolution: Selected Papers,* ed. W. B. Provine. Chicago: Chicago University Press.

———. [1932] 1986. The roles of mutation, inbreeding, crossbreeding, and selection in evolution. In *Evolution: Selected Papers,* ed. W. B. Provine. Chicago: Chicago University Press.

9. Adaptation in Action

Cave-Browne, J. 1857. *Indian Infanticide: Its Origin, Progress, and Suppression.* London: W. H. Allen.

Chambers, G. K. 1988. The *Drosophila* alcohol dehydrogenase gene-enzyme system. In *Advances in Genetics,* 25, eds. E. W. Caspari and J. G. Scandalois. New York: Academic Press.

Charlesworth, B. 1994. *Evolution in Age Structured Populations.* Cambridge: Cambridge University Press.

Davies, N. B. 1992. *Dunnock Behaviour and Social Evolution.* Oxford: Oxford University Press.

Dawkins, R. 1976. *The Selfish Gene.* Oxford: Oxford University Press.

De Buffrénil, V., J. O. Farlow, and A. de Ricqulès. 1986. Growth and function of *Stegosaurus* plates: evidence from bone histology. *Paleobiology* 12: 459–473.

Dickemann, M. 1979. Female infanticide and the reproductive strategies of stratified human societies. In *Evolutionary Societies and Human Social Behavior,* eds. N. A. Chagnon and W. Irons. North Scituate: Duxbury.

Farlow, J. O., C. V. Thompson, and D. E. Rosner. 1976. Plates of the dinosaur Stegosaurus: forced convection heat loss fins? *Science* 192: 1123–1125.

Fastovsky, D. E., and D. B. Weishampel. 1996. *The Evolution and Extinction of the Dinosaurs.* Cambridge: Cambridge University Press.

Freriksen, A., B. L. A. De Ruiter, H.-J. Groenenberg, W. Scharloo, and P. W. H. Henstra. 1994. A multilevel approach to the significance of genetic variation in alcohol dehydrogenase of *Drosophila. Evolution* 48: 781–790.

Gibson, H. L., T. W. May, and A. V. Wilks. 1981. Genetic variation at the alcohol dehydrogenase locus in *Drosophila melanogaster* in relation to environmental variation: ethanol levels in breeding sites and allozyme frequencies. *Oecologia* 51: 191–198.

Gosling, L. M. 1986. Selective abortion of entire litters in the coypu: adaptive control of offspring production in relation to quality and sex. *American Naturalist* 127: 772–795.

Hamilton, W. D. 1964a. The genetical evolution of social behaviour, 1. *Journal of Theoretical Biology* 7: 1–16.

———. 1964b. The genetical evolution of social behaviour, 2. *Journal of Theoretical Biology* 7: 17–32.

———. 1975. Innate social aptitudes of man: an approach from evolutionary genetics. In *ASA Studies, 4: Biosocial Anthropology,* ed. R. Fox. London: Malaby.

Hrdy, S. B. 1999. *Mother Nature: A History of Mothers, Infants, and Natural Selection*. New York: Pantheon Books.

Lewontin, R. C. 1974. *The Genetic Basis of Evolutionary Change.* New York: Columbia University Press.

Maynard Smith, J. 1982. *Evolution and the Theory of Games.* Cambridge: Cambridge University Press.

McDonald, J. H., G. K. Chambers, J. David, and F. J. Ayala. 1977. Adaptive response due to changes in gene regulation: a study with *Drosophila. Proceedings of the National Academy of Sciences USA* 74: 4562–4566.

McDonald, J. H., and M. Kreitman. 1991. Adaptive protein evolution at the Adh locus in *Drosophila. Nature* 351: 652–654.

Mercot, H., D. Defaye, P. Capy, E. Pla, and J. R. David. 1994. Alcohol tolerance, ADH activity, and ecological niche of *Drosophila* species. *Evolution* 48: 756–757.

Nesse, R., and G. C. Williams. 1995. *Evolution and Healing: The New Science of Darwinian Medicine.* London: Weidenfeld and Nicolson.

Oakeshott, J. G., T. W. May, J. B. Gibson, and D. A. Willcocks. 1982. Resource partitioning in five domestic *Drosophila* species and its relationship to ethanol metabolism. *Australian Journal of Zoology* 30: 547–556.

Reznick, D. N., and J. A. Endler. 1982. The impact of predation on life history evolution in Trinidadian guppies (*Poecilia reticulata*). *Evolution* 36: 160–177.

Reznick, D. N., and J. Travis. 1996. The empirical study of adaptation in natural populations. In *Adaptation,* eds. M. R. Rose and G. V. Lauder. San Diego: Academic Press.

Trivers, R. L., and D. E. Willard. 1973. Natural selection of parental ability to vary the sex ratio of offspring. *Science* 179: 90–92.

Williams, G. C. 1966. *Adaptation and Natural Selection.* Princeton: Princeton University Press.

Wilson, E. O. 1978. *On Human Nature.* Cambridge: Cambridge University Press.

Wynne-Edwards, V. C. 1962. *Animal Dispersion in Relation to Social Behaviour.* Edinburgh: Oliver and Boyd.

10. *Theory and Test*

Akashi, H. 1994. Synonymous codon usage in *Drosophila melanogaster*: natural selection and translational accuracy. *Genetics* 144: 927–935.

Bakker, R. T. 1983. The deer flees, the wolf pursues: incongruencies in predator-prey coevolution. In *Coevolution*, ed. D. J. Futuyma and M. Slatkin. Sunderland: Sinauer.

Barrett, P. H., P. J. Gautrey, S. Herbert, D. Kohn, and S. Smith, eds. 1987. *Charles Darwin's Notebooks, 1836–1844*. Ithaca: Cornell University Press.

Bennett, J. H., ed. 1983. *Natural Selection, Heredity, and Eugenics. Including Selected Correspondence of R. A. Fisher with Leonard Darwin and Others*. Oxford: Oxford University Press.

Berry, R. J. 1986. Genetics of insular populations of mammals, with particular reference to differentiation and founder effects in British small mammals. *Biological Journal of the Linnaean Society* 28: 205–230.

Brandon, R. N., and M. D. Rauscher. 1996. Testing adaptationism: a comment on Orzack and Sober. *American Naturalist* 148: 189–201.

Buri, P. 1956. Gene frequencies in small populations of mutant *Drosophila*. *Evolution* 10: 367–402.

Coyne, J. A., N. H. Barton, and M. Turelli. 1997. Perspective: a critique of Sewall Wright's shifting balance theory of evolution. *Evolution* 51: 643–671.

Darwin, C. 1871. *The Descent of Man*. London: John Murray.

———. 1959. *The Origin of Species by Charles Darwin: A Variorum Text*. Ed. M. Peckham. Philadelphia: University of Pennsylvania Press.

Davies, N. B. 1992. *Dunnock Behaviour and Social Evolution*. Oxford: Oxford University Press.

Dawkins, R. 1982. *The Extended Phenotype: The Gene as the Unit of Selection*. Oxford: Freeman.

———. 1986. *The Blind Watchmaker*. New York: Norton.

Freeman, S., and J. C. Herron. 1998. *Evolutionary Analysis*. Englewood Cliffs: Prentice-Hall.

Gould, S. J. 1974. The evolutionary significance of "bizarre" structures: antler size and skull size in the "Irish Elk," *Megalocerus giganteus*. *Evolution* 28: 191–220.

———. 1988. On replacing the idea of progress with an operational notion of directionality. In *Evolutionary Progress*, ed. M. H. Nitecki. Chicago: University of Chicago Press.

———. 1989. *Wonderful Life: The Burgess Shale and the Nature of History*. New York: Norton.

———. 1996. *Full House: The Spread of Excellence from Plato to Darwin*. New York: Paragon.

———. 2002. *The Structure of Evolutionary Theory.* Cambridge: Harvard University Press.

Gould, S. J., and E. A. Lloyd. 1999. Individuality and adaptation across levels of selection: how shall we name and generalize the unit of Darwinism? *Proceedings of the National Academy of Sciences USA* 96: 11904–11909.

Gould, S. J., D. M. Raup, J. J. Sepkoski Jr., T. J. M. Schopf, and D. M. Simberloff. 1977. The shape of evolution: a comparison of real and random clades. *Paleobiology* 3: 23–40.

Hamilton, W. D. 1967. Extraordinary sex ratios. *Science* 156: 477–488.

Hamilton, W. D., R. Axelrod, and R. Tanese. 1990. Sexual reproduction as an adaptation to resist parasites. *Proceedings of the National Academy of Sciences, USA* 87: 3566–3573.

Harcourt, A. H., P. H. Harvey, S. G. Larson, and R. V. Short. 1981. Testis weight, body weight and breeding system in primates. *Nature* 293: 55–57.

Harvey, P., and R. May. 1989. Out for the sperm count. *Nature* 337: 508–509.

Harvey, P., and M. D. Pagel. 1991. *The Comparative Method in Evolutionary Biology.* Oxford: Oxford University Press.

Herre, E. A., C. A. Machado, and S. A. West. 2001. Selective regime and fig wasp sex ratios: toward sorting rigor from pseudo-rigor in tests of adaptation. In *Adaptation and Optimality,* eds. S. H. Orzack, and E. Sober. Cambridge: Cambridge University Press.

Herre, E. A., S. A. West, J. M. Cook, S. G. Compton, and F. Kjellberg. 1997. Fig wasp mating systems: pollinators and parasites, sex ratio adjustment and male polymorphism, population structure and its consequences. In *Social Competition and Cooperation in Insects and Arachnids,* 1: *The Evolution of Mating Systems,* eds. J. Choe and B. Crespi. Cambridge: Cambridge University Press.

Hölldobler, B., and E. O. Wilson. 1990. *The Ants.* Cambridge: Harvard University Press.

Hull, D. L. 1980. Individuality and selection. *Annual Review of Ecology and Systematics* 11: 311–332.

Huxley, J. S. 1942. *Evolution: The Modern Synthesis.* London: Allen and Unwin.

Kimura, M. 1983. *Neutral Theory of Molecular Evolution.* Cambridge: Cambridge University Press.

Lewontin, R. C. 1978. Adaptation. *Scientific American* 239: 176–193.

Lloyd, E. A., and S. J. Gould. 1993. Species selection on variability. *Proceedings of the National Academy of Sciences* 90: 595–599.

Maynard Smith, J. 1978a. *The Evolution of Sex.* Cambridge: Cambridge University Press.

———. 1978b. Optimization theory in evolution. *Annual Review of Ecology and Systematics* 9: 31–56.

Mayr, E. 1954. Change of genetic environment and evolution. In *Evolution as a*

Process, eds. J. S. Huxley, A. C. Hardy, and E. B. Ford. London: Allen and Unwin.

McShea, D. W. 1991. Complexity and evolution: what everybody knows. *Biology and Philosophy* 6: 303–325.

Orzack, S. H., and E. Sober. 1994a. How (not) to test an optimality model. *Trends in Ecology and Evolution* 9: 265–267.

———. 1994b. Optimality models and the test of adaptationism. *American Naturalist* 143: 361–380.

Oster, G., and E. O. Wilson. 1978. *Caste and Ecology in the Social Insects.* Princeton: Princeton University Press.

Raup, D. M. 1986. *The Nemesis Affair: A Story of the Death of Dinosaurs and the Ways of Biology.* New York: Norton.

———. 1988. Testing the fossil record for evolutionary progress. In *Evolutionary Progress,* ed. M. H. Nitecki. Chicago: University of Chicago Press.

Raup, D. M., and S. M. Stanley. 1978. *Principles of Paleontology,* 2nd ed. San Francisco: Freeman.

Reeve, H. K., and L. Keller. 1999. Levels of selection: burying the units-of-selection debate and unearthing the crucial new issues. In *Levels of Selection in Evolution,* ed. L. Keller. Princeton: Princeton University Press.

Ruse, M. 1993. Evolution and progress. *Trends in Ecology and Evolution* 8: 55–59.

Sepkoski J. J., Jr. 1978. A kinetic model of Phanerozoic taxonomic diversity, I: Analysis of marine orders. *Paleobiology* 4: 223–251.

Sober, E. 1984. *The Nature of Selection.* Cambridge: MIT Press.

Tishkoff, S. A., E. Dietzsch, W. Speed, et al. 1996. Global patterns of linkage disequilibrium at the CD4 locus and modern human origins. *Science* 271: 1380–1387.

Vermeij, G. J. 1987. *Evolution and Escalation.* Princeton: Princeton University Press.

Williams, G. C. 1992. *Natural Selection: Domains, Levels and Challenges.* Oxford: Oxford University Press.

Wilson, E. O. 1992. *The Diversity of Life.* Cambridge: Harvard University Press.

Wynne-Edwards, V. C. 1962. *Animal Dispersion in Relation to Social Behaviour.* Edinburgh: Oliver and Boyd.

11. Formalism Redux

Alberch, P. 1982. Developmental constraints in evolutionary processes. In *Evolution and Development,* ed. J. T. Bonner. New York: Springer-Verlag.

Amundson, R. 1994. Two concepts of constraint: adaptationism and the challenge from developmental biology. *Philosophy of Science* 61: 556–578.

Ayala, F. J. 1987. The biological roots of morality. *Biology and Philosophy* 2: 235–252.

Balter, M. 2002. What made humans modern? *Science* 295: 1219–1225.

Carroll, S. B., J. K. Grenier, and S. D. Weatherbee. 2001. *From DNA to Diversity: Molecular Genetics and the Evolution of Animal Design*. Oxford: Blackwell.

Cavalli-Sforza, L. L., and W. F. Bodmer. 1971. *The Genetics of Human Populations*. San Francisco: Freeman.

Charlesworth, B., R. Lande, and M. Slatkin. 1982. A neo-Darwinian commentary on macroevolution. *Evolution* 36: 474–498.

Coates, M. I., and J. A. Clack. 1990. Polydactyly in the earliest tetrapod limbs. *Nature* 347: 66–69.

Dawkins, R. 1986. *The Blind Watchmaker*. New York: Norton.

Douady, S., and Y. Couder. 1992. Phyllotaxis as a physical self-organised growth process. *Physical Review Letters* 68: 2098–2101.

Eldredge, N., and S. J. Gould. 1972. Punctuated equilibria: an alternative to phyletic gradualism. In *Models in Paleobiology*, ed. T. J. M. Schopf. San Francisco: Freeman.

Futuyma, D. J. 1987. On the role of species in anagenesis. *American Naturalist* 130: 465–473.

Ghiselin, M. T. 1980. The failure of morphology to assimilate Darwinism. In *The Evolutionary Synthesis: Perspectives on the Unification of Biology*, eds. E. Mayr and W. Provine. Cambridge: Harvard University Press.

Gilbert, S. F., J. M. Opitz, and R. A. Raff. 1996. Resynthesizing evolutionary and developmental biology. *Developmental Biology* 173: 357–372.

Goodwin, B. 2001. *How the Leopard Changed Its Spots*, 2nd ed. Princeton: Princeton University Press.

Gould, S. J. 1971. D'Arcy Thompson and the science of form. *New Literary History* 2: 229–258.

Gould, S. J., and R. C. Lewontin. 1979. The spandrels of San Marco and the Panglossian paradigm: a critique of the adaptationist program. *Proceedings of the Royal Society of London, Series B: Biological Sciences* 205: 581–598.

Gray, A. 1881. *Structural Botany*, 6th ed. London: Macmillan.

Kauffman, S. A. 1995. *At Home in the Universe: The Search for the Laws of Self-Organization and Complexity*. New York: Oxford University Press.

Lewontin, R. C. 1991. *Biology as Ideology: The Doctrine of DNA*. Toronto: Anansi.

Maynard Smith, J. 1981. Did Darwin get it right? *London Review of Books* 3, no. 11: 10–11.

Maynard Smith, J. R. Burian, S. Kauffman, P. Alberch, J. Campbell, B. Goodwin, R. Lande, D. Raup, and L. Wolpert. 1985. Developmental constraints and evolution. *Quarterly Review of Biology* 60: 265–287.

Mitchison, G. J. 1977. Phyllotaxis and the Fibonacci series. *Science* 196: 270–275.

Niklas, K. J. 1988. The role of phyllotactic pattern as a "developmental constraint" on the interception of light by leaf surfaces. *Evolution* 42: 1–16.

Raff, R. 1996. *The Shape of Life: Genes, Development, and the Evolution of Animal Form*. Chicago: University of Chicago Press.

Reeve, H. K., and P. W. Sherman. 1993. Adaptation and the goals of evolutionary research. *Quarterly Review of Biology* 68: 1–32.

Ruse, M. 1998. *Taking Darwin Seriously: A Naturalistic Approach to Philosophy*, 2nd ed. Buffalo: Prometheus.

Ruse, M., and E. O. Wilson. 1986. Moral philosophy as applied science. *Philosophy* 61: 173–192.

Russell, E. S. 1916. *Form and Function: A Contribution to the History of Animal Morphology*. London: John Murray

Thompson, D. W. 1948. *On Growth and Form*, 2nd ed. Cambridge: Cambridge University Press.

Vogel, S. 1988. *Life's Devices: The Physical World of Animals and Plants*. Princeton: Princeton University Press.

Wagner, G. P. 1988. The influence of variation and of developmental constraints on the rate of multivariate phenotypic evolution. *Journal of Evolutionary Biology* 1: 45–66.

Williams, G. C. 1992. *Natural Selection: Domains, Levels and Challenges*. Oxford: Oxford University Press.

Wright, C. 1871. *Darwinism: Being an Examination of Mr. St. George Mivart's "Genesis of Species."* London: John Murray.

12. From Function to Design

Amundson, R., and G. V. Lauder. 1994. Function without purpose: the uses of causal role function in evolutionary biology. *Biology and Philosophy* 9: 443–469.

Bergson, H. 1911. *Creative Evolution*. New York: Holt.

Black, M. 1962. *Models and Metaphors*. Ithaca: Cornell University Press.

Clarkson, E. N. K., and R. Levi-Setti. 1975. Trilobite eyes and the optics of Descartes and Huygens. *Nature* 254: 663–667.

Cooper, J. M., ed. 1997. *Plato: Complete Works*. Indianapolis: Hackett.

Cummins, R. 1975. Functional analysis. *Journal of Philosophy* 72: 741–764.

Huxley, J. S. 1912. *The Individual in the Animal Kingdom*. Cambridge: Cambridge University Press.

Kant, I. [1790] 1928. *The Critique of Teleological Judgement*. Trans. J. C. Meredith. Oxford: Oxford University Press.

Levi-Setti, R. 1993. *Trilobites*. Chicago: University of Chicago Press.

Mayr, E. 1988. *Toward a New Philosophy of Biology: Observations of an Evolutionist*. Cambridge: Harvard University Press.

Nagel, E. 1961. *The Structure of Science: Problems in the Logic of Scientific Explanation*. New York: Harcourt, Brace and World.

Rudwick, M. J. S. 1964. The inference of function from structure in fossils. *British Journal for the Philosophy of Science* 15: 27–40.

Simpson, G. G. 1949. *The Meaning of Evolution*. New Haven: Yale University Press.

Sommerhoff, G. 1950. *Analytical Biology*. Oxford: Oxford University Press.

Waddington, C. H. 1957. *The Strategy of the Genes*. London: Allen and Unwin.

Williams, G. C. 1966. *Adaptation and Natural Selection*. Princeton: Princeton University Press.

Wright, L. 1973. Functions. *Philosophical Review* 82: 139–168.

13. *Design as Metaphor*

Allen, C., and M. Bekoff. 1995. Biological function, adaptation, and natural design. *Philosophy of Science* 62: 609–622.

Beckner, M. 1969. Function and teleology. *Journal of the History of Biology* 2: 151–164.

Darwin, C. 1862. *On the Various Contrivances by Which British and Foreign Orchids Are Fertilized by Insects, and on the Good Effects of Intercrossing*. London: John Murray.

———. 1868. *The Variation of Animals and Plants under Domestication*. London: John Murray.

Dawkins, R. 1983. Universal Darwinism. In *Molecules to Men*, ed. D. S. Bendall. Cambridge: University of Cambridge Press.

De Waal, F. 1982. *Chimpanzee Politics: Power and Sex among Apes*. London: Cape.

Fodor, J. 1996. Peacocking. *London Review of Books*, April, 19–20.

Ghiselin, M. T. 1983. Lloyd Morgan's canon in evolutionary context. *Behavioral and Brain Sciences* 6: 362–363.

Godfrey-Smith, P. 1999. Adaptationism and the power of selection. *Biology and Philosophy* 14: 181–194.

———. 2001. Three kinds of adaptationism. In *Adaptationism and Optimality*, eds. S. H. Orzack and E. Sober. Cambridge: Cambridge University Press.

Goodall, J. 1986. *The Chimpanzees of Gombe: Patterns of Behavior*. Cambridge: Harvard University Press.

Hesse, M. 1966. *Models and Analogies in Science*. Notre Dame: University of Notre Dame Press.

Huxley, L. 1900. *The Life and Letters of Thomas Henry Huxley*. London: Macmillan.

Kant, I. [1790] 1928. *The Critique of Teleological Judgement*. Trans. J. C. Meredith. Oxford: Oxford University Press.

Maynard Smith, J. 1969. The status of neo-Darwinism. In *Towards a Theoretical Biology*, ed. C. H. Waddington. Edinburgh: Edinburgh University Press.

Mayr, E. 1988. *Toward a New Philosophy of Biology: Observations of an Evolutionist*. Cambridge: Harvard University Press.

McLaughlin, P. 2001. *What Functions Explain: Functional Explanation and Self-Reproducing Systems.* Cambridge: University of Cambridge Press.

Nagel, E. 1961. *The Structure of Science, Problems in the Logic of Scientific Explanation.* New York: Harcourt, Brace and World.

Witmer, L. M. 2001. Nostril position in dinosaurs and other vertebrates and its significance for function. *Science* 293: 850–853.

14. Natural Theology Evolves

Argyll, Duke of. 1867. *The Reign of Law.* London: Alexander Strahan.

Barth, K. 1933. *The Epistle to the Romans.* Oxford: Oxford University Press.

Bryan, W. J. 1922. *In His Image.* New York: Fleming H. Revell.

Butler, J. [1736] 1824. *The Analogy of Religion, Natural and Revealed, to the Constitution and Course of Nature.* London: Longman, Hurst, Rees, Orme.

Ellegård, A. 1958. *Darwin and the General Reader.* Goteborg: Goteborgs Universitets Arsskrift.

Hodge, C. 1872. *Systematic Theology.* London: Nelson.

John Paul II. 1997. The Pope's message on evolution. *Quarterly Review of Biology* 72: 377–383.

———. 1998. *Fides et Ratio: Encyclical Letter of John Paul II to the Catholic Bishops of the World.* Vatican City: L'Osservatore Romano.

Larson, E. J. 1997. *Summer for the Gods: The Scopes Trial and America's Continuing Debate over Science and Religion.* New York: Basic Books.

Livingstone, D. N. 1987. *Darwin's Forgotten Defenders: The Encounter between Evangelical Theology and Evolutionary Thought.* Grand Rapids: Eerdmans.

McCosh, J. 1871. *Christianity and Positivism: A Series of Lectures to the Times on Natural Theology and Christian Apologetics.* London: Macmillan.

———. 1887. *Realistic Philosophy Defended in a Philosophic Series.* New York.

McMullin, E. 1985. Introduction: evolution and creation. In *Evolution and Creation,* ed. E. McMullin. Notre Dame: University of Notre Dame Press.

Mivart, St. G. 1871. *On the Genesis of Species.* London: Macmillan.

———. 1876. *Contemporary Evolution.* London: Henry S. King.

Newman, J. H. [1845] 1989. *An Essay on the Development of Christian Doctrine.* Notre Dame: University of Notre Dame Press.

———. 1971. *The Letters and Diaries of John Henry Newman,* 21. Eds. C. S. Dessain and T. Gornall. Oxford: Clarendon Press.

———. 1973. *The Letters and Diaries of John Henry Newman,* 25. Eds. C. S. Dessain and T. Gornall. Oxford: Clarendon Press.

Numbers, R. L. 1998. *Darwinism Comes to America.* Cambridge: Harvard University Press.

Numbers, R. L., and J. Stenhouse, eds. 1999. *Disseminating Darwinism: The Role of Place, Race, Religion, and Gender.* Cambridge: Cambridge University Press.

Pannenberg, W. 1993. *Towards a Theology of Nature.* Louisville: Westminster/John Knox Press.

Plantinga, A. 1983. Reason and belief in God. In *Faith and Rationality: Reason and Belief in God,* eds. A. Plantinga and N. Wolterstorff. Notre Dame: University of Notre Dame Press.

Powell, B. 1860. On the study of the evidences of Christianity. In *Essays and Reviews.* London: Longman, Green, Longman, and Roberts.

Raven, C. E. 1943. *Science, Religion, and the Future.* Cambridge: Cambridge University Press.

———. 1953. *Natural Religion and Christian Theology: Experience and Interpretation.* Cambridge: Cambridge University Press.

Rolston H., III. 1987. *Science and Religion.* New York: Random House.

Ruse, M. 2000. *The Evolution Wars: A Guide to the Controversies.* Santa Barbara, California: ABC-CLIO.

Sedgwick, A. [1860] 1988. Objections to Mr Darwin's theory of the origin of species. In *But Is It Science? The Philosophical Question in the Creation/Evolution Controversy,* ed. M. Ruse. Buffalo: Prometheus.

Teilhard de Chardin, P. 1959. *The Phenomenon of Man.* London: Collins.

Temple, F. 1884. *The Relations between Religion and Science.* London: Macmillan.

Tennant, F. R. 1930. *Philosophical Theology,* 2: *The World, the Soul, and God.* Cambridge: Cambridge University Press.

15. Turning Back the Clock

Behe, M. 1996. *Darwin's Black Box: The Biochemical Challenge to Evolution.* New York: Free Press.

Darwin, C. 1859. *On the Origin of Species.* London: John Murray.

Dawkins, R. 1983. Universal Darwinism. In *Molecules to Men,* ed. D. S. Bendall. Cambridge: University of Cambridge Press.

———. 1986. *The Blind Watchmaker.* New York: Norton.

———. 1995. *A River Out of Eden.* New York: Basic Books.

———. 1997. Obscurantism to the rescue. *Quarterly Review of Biology* 72: 397–399.

Dembski, W. A. 1998. *The Design Inference: Eliminating Chance through Small Probabilities.* Cambridge: Cambridge University Press.

———. 2000. The third mode of explanation: detecting evidence of intelligent design in the sciences. In *Science and Evidence for Design in the Universe,* eds. M. J. Behe, W. A. Dembski, and S. C. Meyer. San Francisco: Ignatius Press.

———. 2001. What every theologian should know about creation, evolution and design. In *Unapologetic Apologetics: Meeting the Challenges of Theological Studies,* eds. W. A. Dembski and W. J. Richards. Downers Grove, IL: InterVarsity Press.

———. 2002. *No Free Lunch: Why Specified Complexity Cannot Be Purchased without Intelligence.* Lanham, MD: Rowman & Littlefield.

Jacob, F. 1977. Evolution and tinkering. *Science* 196: 1161–1166.

Johnson, P. E. 1991. *Darwin on Trial.* Washington, DC: Regnery Gateway.

Meléndez-Hevia, E., T. G. Waddell, and M. Cascante. 1996. The puzzle of the Krebs citric acid cycle: assembling the pieces of chemically feasible reactions, and opportunism in the design of metabolic pathways during evolution. *Journal of Molecular Evolution* 43: 293–303.

Miller, K. 1999. *Finding Darwin's God.* New York: Harper and Row.

Morris, H. M., et al. 1974. *Scientific Creationism.* San Diego: Creation-Life Publishers.

Pannenberg, W. 1993. *Towards a Theology of Nature.* Louisville: Westminster/John Knox Press.

Raven, C. E. 1953. *Natural Religion and Christian Theology: Experience and Interpretation.* Cambridge: Cambridge University Press.

Ray, T. S. 1996. An approach to the synthesis of life. In *The Philosophy of Artificial Life,* ed. M. A. Boden. Oxford: Oxford University Press.

Reichenbach, B. R. 1982. *Evil and a Good God.* New York: Fordham University Press.

Ruse, M. 2001. *Can a Darwinian Be a Christian? The Relationship between Science and Religion.* Cambridge: Cambridge University Press.

ILLUSTRATION CREDITS

Introduction: Stegosaurus (*Scientific American,* September 1978, 217), by permission of the estate of Bunji Tagawa)

Chapter 1: Praying hands, by Albrecht Dürer, 1508 (Graphische Albertina, Vienna)

Chapter 2: White rabbit looking at watch, by John Tenniel (Lewis Carroll, *Alice in Wonderland,* 1866)

Chapter 3: Tertiary mammal *Palaeotherium* (reconstructed by Georges Cuvier, from William Buckland, Bridgewater Treatise, 1836)

Chapter 4: "Old Bones," by Richard Owen (*Vanity Fair* cartoon, 1873)

Chapter 5: La Belle Hottentot (Hottentot Venus), French cartoon (circa 1814–1815)

Chapter 6: W. Heath Robinson, "The Professor's Invention for Peeling Potatoes" (Norman Hunter, *The Incredible Adventures of Professor Branestawm,* 1933)

Chapter 7: Eohippus and Eohomo, cartoon by Thomas Henry Huxley, circa 1873 (Huxley Archives, London)

Chapter 8: Snails *(Cepaea nemoralis)* (P. M. Sheppard, *Natural Selection and Heredity,* 1958)

Chapter 9: Male dunnock pecking exposed cloaca of female (Nicholas Davies, *Dunnock Behaviour and Social Evolution,* 1992; illustration by David Quinn, by permission)

Chapter 10: Irish elk (reconstructed by Richard Owen, *A History of British Fossil Mammals and Birds,* 1846)

Chapter 11: "Still Life: Vase with Fifteen Sunflowers," by Vincent van Gogh, 1888 (National Gallery, London)

Chapter 12: Trilobite (Richard Fortey, *Fossils: The Key to the Past,* 1983)

Chapter 13: Sauropod dinosaur Diplodocus (L. M. Witmer, "Nostril Position in Dinosaurs and Other Vertebrates and Its Significance for Function," *Science,* 2001, 293, 850, reprinted with permission. Copyright 2001 American Association for the Advancement of Science, painting by Bill Parsons, under the direction of L. M. Witmer)

Chapter 14: "Pedigree of Man" (Ernst Haeckel, *The Evolution of Man,* 1896)

Chapter 15: Mousetraps, by John McDonald, circa 2000 (reproduced with permission)

ACKNOWLEDGMENTS

As always, I am much in debt to foundations, private and public, who have supported me in my work, to the institutions to which I have belonged and those where I have spent times of leave, to libraries that have given me access to their holdings, printed and archival, and most of all to individual people—those at my home universities, librarians, archivists, and my fellow scholars, all of whom have been helpful and suggestive and some of whom have been convinced.

I am the recipient of a generous grant from the John Marks Templeton Foundation, whose support has been of specific value in preparing this book. For many years I have gratefully accepted funding from the (Canadian) Social Sciences and Humanities Research Council and, more recently, from the William and Lucyle Werkmeister Foundation at Florida State University. Colleagues who have given me help and advice in the writing of this manuscript include Barry Allen, Michael Cavanaugh, Russ Dancy, Ward Goodenough, Philip Hefner, Ernan McMullin, and Edward T. Oakes, S.J. I am obliged to Richard Dawkins for putting me right on Darwin's thinking about sex ratios. As always, my good friends David L. Hull, Robert J. Richards, and Edward O. Wilson have worried about what I am going to say next on the theory that we all love and cherish. Christopher Pynes and Jeremy Kirby have worked hard as my research assistants. At Harvard University Press, Michael Fisher has been all one could want in an editor and much more. And once again Susan Wallace Boehmer has been a wonderful manuscript editor. Simply the best I have ever had.

My family—Lizzie, Emily, Oliver, and Edward—are more important to me than anything.

INDEX